CHEMISTRY
FOR THE
PROTECTION
OF THE
ENVIRONMENT 4

ENVIRONMENTAL SCIENCE RESEARCH

Series Edtior:

Herbert S. Rosenkranz

Department of Environmental and Occupational Health
Graduate School of Public Health
University of Pittsburgh
130 DeSoto Street
Pittsburgh, Pennsylvania

Founding Editor:

Alexander Hollaender

A Continuation Order Plan is available for this series. A continuation order will bring delivery of each new
volume immediately upon publication. Volumes are billed only upon actual shipment. For further information
please contact the publisher.

Preface

The central goal of Chemistry for the Protection of the Environment conference series is to improve technology transfer and scientific dialogue, thereby leading to a better comprehension of and solution to a broad spectrum of environmentally related problems.

The first meeting in the CPE series was organized by Professor Lucjan Pawlowski and Dr. William Lacy in 1976 at the Marie Curie-Sklodowska University in Lublin, Poland. The conference dealt with various physicochemical methodologies for water and wastewater treatment research projects that were jointly sponsored by the United State Environmental Protection Agency (EPA) and Poland.

The great interest expressed by the participants led the organizers to expand the scope of the second conference, which was also held in Poland in September 1979. A third and enlarged symposium was again successfully held in Lublin, Poland in 1981. At that time, the participating scientists and engineers expressed their desire to broaden the coverage as well as the title "Chemistry for the Protection of the Environment." The next meeting, CPE IV, was convened in September 1983 at the Paul Sabatier University in Toulouse, France, and included participants from various government agencies, academia, and the private sector, representing industrialized countries as well as emerging nations from both the East and West, in an independent non-political forum.

CPE V, held in September 1985 at the Catholic University in Leuven, Belgium, covered topics dealing with treatment technologies and phenomena related to hazardous waste and the utilization of fossil fuels. It provided an opportunity for interdisciplinary discussions and encouraged the exchange of ideas among international specialists from diverse fields and backgrounds.

CPE VI was held in 1987 at the University of Turin in Italy, with over 150 scientific papers and posters presented to an audience from 32 nations. This assemblage comprised in equal measure scientists from Europe, the New World, and developing nations.

CPE VII was convened at the Catholic University in Lublin, Poland in 1989. The exchange of information by approximately 200 scientists and engineers made this a memorable scientific meeting. Distinguished participants included Poland's Minister and Deputy Minister for Environmental Protection, U.S. Scientific Council, Israel's Deputy Minister of the Environment, presidents and vice presidents of five universities, representatives of the Academies of Sciences for Czechoslovakia, France, Italy, Poland, and the U.S.S.R., as well as many department heads and acclaimed scientists.

In September 1991, CPE VIII was convened in Lublin, Poland. The technical presentations were original and informative, with the major topics being chemical/physical/biological treatment technologies, monitoring, modeling, and risk assessment.

CPE IX, held in September 1993 in Alexandria, Cairo/Luxor, Egypt, was a joint conference with EPA/U.S. AID's Fourth International Symposium on Industry in the Developing World. This included a workshop on industrial pollution prevention and clean technologies, and other cooperation and institutional issues. Participants were comprised of a multi-disciplined technical group from 27 countries.

CPE X, the 20th anniversary meeting, was held in the city of CPE's birth, Lublin, Poland. Papers and posters on technology transfer; novel, innovative and alternative treatment processes; and environmental problems facing countries were presented. The meeting was enhanced by the participation of large delegations from both The Peoples Republic of China and the Taiwan Chinese Republic.

CPE XI returned to Cairo, Egypt in 1997, and papers and posters were presented on adsorption, analytical methods, chemical/biological/treatment, groundwater studies, ion exchange, modeling, risk assessment, waste minimization and treatment, and for the first time, ISO 14001, which focuses on environmental management and quality systems.

CPE XII took the conference series to the other side of the world, and was held in Nanjing, China in September 1999. Once again, the conference brought environmental scientists, engineers, and policymakers together to present innovative solutions to environmental problems and to develop collaborations.

CPE XIII, marked the 25th anniversary of CPE and the first time the CPE conference was convened in the United States. Environmental scientists, engineers, and policymakers from India, Japan, the Philippines, Poland, Germany, the Netherlands, Russia, Slovakia, and the United States gathered to discuss current issues, including new risk assessment methodologies, innovative analytical and waste management techniques, and emerging environmental security policies. Field excursions led by Hilo scientists to the Hawaii Scientific Drilling Project and the Hawaii Volcanoes National Park, and an ecological tour of local rainforests complemented the lectures and promoted interactions among the participants. For the first time, a student poster session was held, and the Tristan J. DLugosz Memorial Award for excellence in student poster presentations was presented to University of Hawaii at Hilo students Mark Albins, Wiley Evans, Megan Flynn and Rachel Horton for their poster "How Does Hilo Bay Function? A Biogeochemical Snapshot of the Waters and Sediments of the Bay. " CPE co-founder and co-president Dr. William J. Lacy was honored with an award commending him for his vision and role in establishing CPE and promoting scientific dialogue and international cooperation among participating environmental scientists and policymakers from around the world. Through his leadership and knowledge, the goals of CPE have become a reality.

Robert Mournighan, CPE XIII Lead Editor
Marjorie Auyong Gonzalez, CPE XIII Chairperson

Contents

CHEMISTRY
FOR THE
PROTECTION
OF THE
ENVIRONMENT 4

1

REMEDIATION OF CONTAMINATED SITES

STUDY OF THE CLINOPTILOLITE-RICH TUFF-BASED COMPOSITES FOR SOME AQUEOUS ANIONIC SPECIES RECOVERY

Eva Chmielewská, Stanislava Nagyová [2], Mária Reháková [3] and Nina Bogdanchikova[4]
Ecosozology and Physiotactics Department, Faculty of Natural Sciences, Comenius University, Bratislava, Slovak Republic, E-mail: chmielewska@fns.uniba.sk; [2]Department of Physics, Electrotechnical Faculty, Technical University, Košice, Slovak Republic; [3]Department of Chemistry, Faculty of Science, University of P. J. Šafarik, Košice, Slovak Republic; [4] Centro de Ciencias de la Materia Condensada, Ensenada, B.C., Mexico

Abstract: Batch and column sorption experiments have been performed to study a significantly enhanced removal of toxic oxyanions, i.e. chromate and arsenate from aqueous effluents on inland clinoptilolite modified by the octadecylammonium acetate [ODA]. The arrangements (orientation and length) of the surface-attached organic chains and thus the initial concentration of ODA – modifier by composite preparation have been the important factor for the difference in the adsorption states of the guest species [oxyanions]. Novel nano-structure inorganic – organic composites prepared have been investigated using SEM, thermogravimetry, NMR, IR, HR TEM, UV-VIS DRS and powder XRD spectral analytical methods. Finally, adsorption isotherms for system studied have been expressed. Some approach for regeneration of exhausted surfactant-immobilized clinoptilolite with inorganic salt solutions under dynamic regime has been proposed, respectively.

Key words: Clinoptilolite, ODA-modifier, chromate and arsenate removal, adsorption, isotherm, and breakthrough curve.

1.   INTRODUCTION

Both arsenic and chromium are considered to pose risks to public health and, in the inorganic state, especially chromium in hexavalent form, are generally highly toxic to biota. Their toxicity to humans was evidenced by many acute poisoning episodes in past; e.g., Nippon Chemical Ind. has apparently been employed in construction materials principal source of airborne chromium, therefore the local groundwater was known to be contaminated.[1] Contamination of drinking water by arsenic has been of concern in Taiwan, Argentina and Chile. In Taiwan, an increased incidence of skin cancer has resulted in the local population and isolated cases of Blackfoot disease have also occurred due to this pollution.[2] Arsenic has been associated with poisoning of humans in the USA through the consumption of "moonshine" whisky, respectively.[3]

In former Czechoslovakia, arsenic emissions from a coal-fired plant are considered to have given rise to respiratory problems. Currently, in Slovakia approaching the environmental legislation of European Union, the State Regulatory Authorities implement the increasingly stringent water quality standards. Consequently to arsenic occurrence in some drinking water reservoirs in the region, some treatment methods for removal of arsenic and chromium from waters have been examined using the naturally abundant domestic clinoptilolite.[4] Arsenic appears mostly in natural air-saturated aquifers in the form of As(V)-anions. However, natural occurrence of the element in underground water bodies is rarely, especially sulfuric ore drainage use to cause its enhanced, over-limit concentration in some artesian wells of the country.

Generally, inorganic arsenic can occur in the environment in several forms but in natural waters and thus in drinking water, it is mostly found as trivalent arsenite or pentavalent arsenate. Organic arsenic species, abundant in seafood, are very much less harmful to health and are readily eliminated by the body [5].

Chromium hexavalent, the more toxic than the tri-valent form, comes in surface waters from anneries, chemical, energetic and ceramic industries. The stable chromate form predominates in alkali, whereas dichromate in acidic waste waters, mostly in deficiency of reductives.[6]

Organo-modified natural zeolites as new tailored natural materials for removal of cations, anions and even organic pollutants may present fairly large potential for water utility companies. The topic of this study was to examine the oxyanions removal from waters by octadecylammonium-enriched inland clinoptilolite. The 18-carbon chain consisting surfactant attached on the clinoptilolite surface, as to the organic acids of living bodies comparable substances, makes the treatment process economic on scale and cost-effective as well.[7]

2. EXPERIMENTAL

2.1 Sample Preparation

Clinoptilolite tuff used in these experiments was supplied by the Mining and Benefication Company of natural zeolite in Slovakia Zeocem Bystré (industrial quarry at Nižný Hrabovec). Grain-size fraction of the sample package was (0.3 – 1.0) mm. Cation exchange capacity (CEC) towards ammonium measured according to Kozáč [8] reached 1.4 mmol/g. Physico-chemical and mineralogical specifications of zeolite rock are reported elsewhere.[9]

Since sorption behaviors of organo-modified zeolite is determined by the properties of both the size of organo-modifier and CEC of zeolite, respecting the economy as well, we chose octadecylammonium (ODA) acetate for the organo-inorganic composite preparation. ODA adsorption was reacted with clinoptilolite by adding the weighed adsorbent to an acidified aqueous adsorbate solution (Merck, AR Grade) under continuous stirring for 24 hr at a temperature of 80 °C. Thereafter, the surfactant-loaded sample was filtered, repeatedly washed with distilled water, oven-dried at 105 °C and ground with a mortar and pestle at the laboratory, respectively. Then such a prepared zeolite was screened mechanically to make the selected particle-size distribution of (0.2 – 0.6) mm. To compare chromate and arsenate removal on organically and inorganically modified clinoptilolite as well, we utilize Ag^+- and Pb^{2+}- exchanged forms of natural zeolite, supposing these will create on the zeolite surface chemically bound precipitates of silver or lead arsenates and chromates. The crushed, raw zeolite samples were loaded up with 4% silver nitrate or lead acetate aqueous solutions following an conventional ion exchange procedure, analogous to that reported by many other authors.[10]

2.2 Apparent Equilibrium and Feasibility Studies

A 0.5-gram mass of either the organo-treated or inorganic cation exchanged zeolite and 50 mL of 10 mM/L arsenate or chromate aqueous solutions were placed into Erlenmeyer flasks and mechanically shaken in reciprocating mode to attain equilibrium. Different equilibrium periods for individual zeolite modifications and both aqueous oxyanions species have been established. The adsorption isotherm experiments were conducted using above mass/ volume ratio of samples with an initial metal concentrations ranged from 0.5 to 100 mM/L at laboratory temperature. The amount of anion sorbed to the zeolite was determined from the difference between the metal concentration in solution before and after equilibrium. Batch desorption experiments have been performed respectively to prove how strongly ODA molecules charged to the zeolite and exchanged cation sites on the surface with the examined oxyanions are bound in various ionic strength solutions. Dynamic fixed bed runs have been carried out in laboratory glass columns by altering the hydraulic down flow load in the range from 10 – 25 Bed Volumes per hour (BV/hr). The initial metal concentrations in tested solutions were about 0.5 mM/L. The operation cycle, until zeolite bed saturation, from 6 to 15 hr., depending on feed flow rate. Furthermore, optimization of regeneration for laboratory columns by reverse flow with several 2% aqueous salt solutions (NaCl, NaNO$_3$, Na$_2$SO$_4$, and Na2CO3) was undertaken. Chemicals for preparation of stock solutions were of analytical grade.

2.3   Analytical Procedures

The zeolite was characterized by means of powder- XRD (Philips – APD computer controlled Diffractometer equipped with APD analytical software, set-up of parameters: CuKα radiation, voltage 30 kV, intensity 15 mA, Co/Ni-filter, diaphragm 1,1,05, 0.1 step, time 1 sec.) and Electron Probe Microanalyser Jeol-JXA 840A (Japan). Ion sputtering device Jeol JFC-1100 was applied for coating of some well-formed single crystals of zeolite rock fragments with a gold alloy after carbon film coating of the surface by High vacuum Balzers BAE 080 device. The concentration of chromates was analysed by means of Diode Array Spectrophotometer Hewlett Packard 8452A (USA) in UV spectral range at wavelength 372 nm using the statistical Quant I program. Arsenates concentration was analysed by means of Atomic Emission Spectrometer with Inductively Coupled Plasma, sequential Plasmakon S35, Kontron (Germany) and Baird ICP 2070 Spectrometer (USA).

3.   RESULTS AND DISCUSSION

X-ray diffraction measurements indicated that the zeolite rock consisted primarily of clinoptilolite (60-70%), volcanic glass (10%), feldspar (10%) and minor quantities of cristobalite, quartz and plagioclase (20%). Fig. 1 represents the XRD pattern of ODA-Clinoptilolite-rich tuff used for arsenate or chromate removal from aqueous solutions.

CHEMISTRY FOR THE PROTECTION OF THE ENVIRONMENT 4

Edited by

Robert Mournighan

U.S. Environmental Protection Agency
Kansas City, Kansas

Marzenna R. Dudzińska

Technical University of Lublin
Lublin, Poland

John Barich

U.S. Environmental Protection Agency
Seattle, Washington

Marjorie Auyong Gonzalez

Lawrence Livermore National Laboratory
Livermore, California

and

Robin Kealoha Black

University of Hawaii at Hilo Conference Center
Hilo, Hawaii

 Springer

A CIP Catalogue record for tis book is available from the Library of Congress.

ISBN 0-387-23020-3   Prined on acid-free paper.

Printed in the United States of America.

9 8 7 6 5 4 3 2 1

springeronline.com

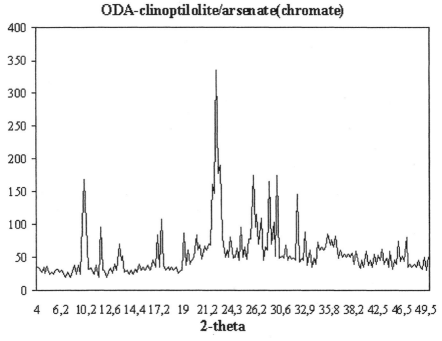

Figure 1. XRD pattern of ODA-modified clinoptilolite-rich tuff adsorbed with arsenate (chromate)

STUDY OF THE CLINOPTILOLITE-RICH TUFF-BASED
COMPOSITES FOR SOME AQUEOUS ANIONIC SPECIES
RECOVERY

Figure 2. Photomicrographs of unmodified, raw; ODA-modified and ODA-modified
clinoptilolite-rich tuff adsorbed with arsenate (from the left to the right); magnification 3700X

However, the polymerous ODA-surfactant attached on the clinoptilolite
surface disabled to appear the characteristic arsenate or chromate peaks at
the X-ray diffractogram except the typical clinoptilolite ones according to
JC PDF 25-1349.

Photomicrographs of the natural (unmodified) clinoptilolite samples
exhibited well-defined, tabular-shaped crystals with excellent crystal edges.
As surface coverage with the polymeric compound occurred, smaller, more
agglomerated crystals and more poorly defined crystal edges were observed
in about 1-μm scale SEM image (Fig. 2). The apparent sharpness of the
images decreased with increasing surfactant coverage.

IR spectra were measured by the KBr disc technique using an Infrared Spectrophotometer 781 (Perkin-Elmer). According to the results of infrared spectroscopy of the raw clinoptilolite-rich tuff (CT) and the ODA- modified two forms (CT ODA 2 and CT ODA 5) all three spectra according to Fig. 3 exhibit a strong broad absorption band in the 1200 - 900 cm^{-1} region, which corresponds to the asymmetric stretching vibrations of the Al-Si-O group. This band changes from a sharp to a broad one, probably when the clinoptilolite-rich tuff becomes amorphous after heating to 900 °C. In the case of CT ODA 2 and CT ODA 5 no significant changes of the absorption band at 1200 - 900 cm^{-1} were observed, except for a slight broadening and decreasing in CT ODA 2. According to these results the ODA in the both forms is bound on the surface. The presence of water in the unmodified sample and the ODA- modified products was observed by two absorption bands. The first one corresponds to the stretching vibrations (3300 - 3600 cm^{-1}) and the second one to the bending vibration of water at 1630 cm^{-1}. However, the intensity of these two bands in CT ODA 2 and CT ODA 5 is lower in comparison to the original material. The presence of $-NH_3^+$ is to check with difficulties due to the overlapping of the absorption bands corresponding to the $-NH_3^+$ and water

STUDY OF THE CLINOPTILOLITE-RICH TUFF-BASED
COMPOSITES FOR SOME AQUEOUS ANIONIC SPECIES
RECOVERY

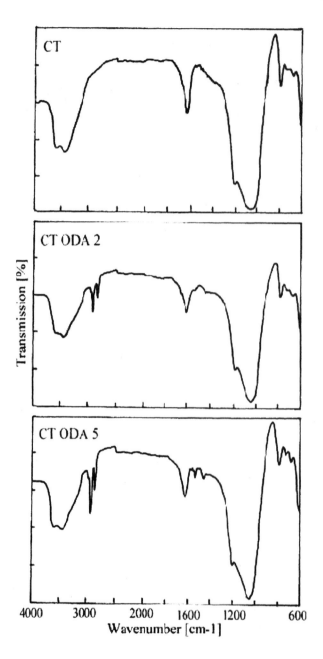

Figure 3. IR spectra of natural clinoptilolite-rich tuff (CT) and the ODA-modified forms (CT ODA 2 and CT ODA 5).

The NMR measurements were carried out at room temperature by using a CW spectrometer constructed at the Department of Physics, Technical University in Kosice with working frequency 10.545 MHz. The attained derivative recordings were averaged out of a few accumulated repetitions and the average was used for the next evaluation of the derivative recordings.

The ^{1}H NMR derivative spectra of natural clinoptilolite sample CT and two modified clinoptilolite samples CT ODA 2 and CT ODA 5 are at Fig. 4. The NMR signal arose from hydrogen in OH groups of water that are placed along the clinoptilolite channel walls, from the CH_3, CH_2 and $-NH_3^+$ groups that are in the ODA chains and also from free water molecules that are always present in natural clinoptilolites.

Consequently to the scope of the NMR measurements to find, how the ODA chains are bound with the clinoptilolite and how they influence the shape of NMR lines, it was necessary to compare the lines with each other and to justify which changes in the structure took place.

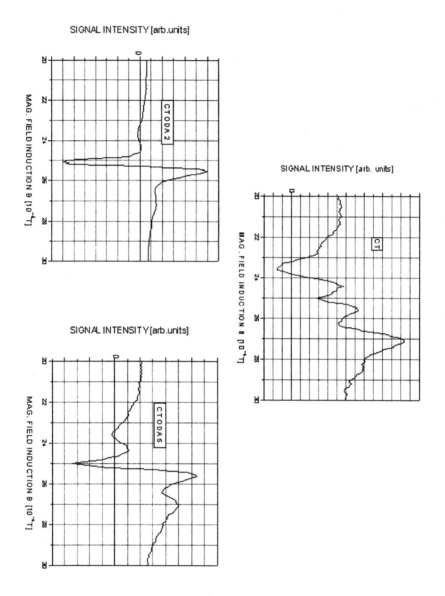

Figure 4. Derivative 1H NMR spectra of the measured clinoptilolite samples at room temperature. The modulation amplitude was 0.22 G, phase detector time constant 0.1s, speed of the field 6.9 mGs -1.The lines are averaged out of eight accumulated repetitions.

For the evaluation of NMR measurements, there are a few very important parameters: second moment M_2 and width of NMR line. They can offer some kind of information about the surrounding of the resonating nuclei, i.e. hydrogen nuclei in that case. It is because the molecules around the resonating nuclei give rise to the magnetic field also in the resonating nuclei place so that they influence the resonance state of the measured hydrogen atoms. The second moment of the experimental lines of CT, CT ODA 2 and CT ODA 5 was calculated from equation published[11] using a computer program. The calculations resulted in the following values of the second moment: M_2=2.02 G^2 for the CT sample, M_2=2.82 G^2 for the CT ODA 2 sample and M_2=5.79 G^2 for the sample CT ODA 5.

The second moment values can also be calculated from the structure. When the nuclei are tightly arranged, i.e. their thermal motion is negligible, M_2 depends on the sixth power of the surrounding nuclei distances. It means, that it is mainly influenced by the nearest atoms around the resonating nuclei. Thus, when the resonating hydrogen nuclei form groups or when there are more of them bound with an element in the sample, M_2 value is higher.

As for the width of NMR resonance lines, it is inversely proportional to mobility of resonating nuclei. The width calculated from the experimental NMR lines of the three samples is of the highest value for the sample CT ODA 5 and there is only a small difference between the linewidths of CT and CT ODA 2.

The above mentioned second moment values of the three samples and their linewidths indicate that the molecules of ODA in the sample CT ODA 5 are more tightly bound in this sample than in the sample CT ODA 2 and that they are placed closer to each other. As for the kind of bond of the ODA within the samples, there is no direct proof about it in the ^{1}H NMR measurements

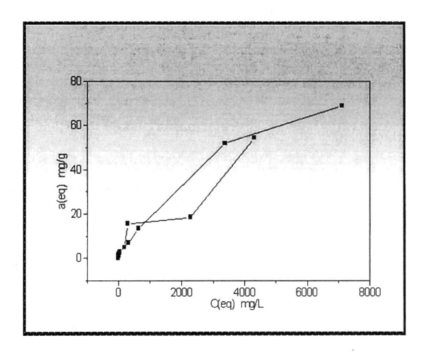

Figure 5. Adsorption isotherms for arsenate (upper line) and chromate (lower line) aqueous solutions with ODA-clinoptilolite, Tconst.= 23°C

The initial ODA concentration for preparation of surfactant modified clinoptilolite was set up according to Haggerty and Bowman[10] to satisfy external CEC of zeolite and to provide a bilayer formation, i.e. attachment of the amine head groups with a permanent 1+ charge to available exchange sites by coulombic interactions and among them upward slipped ODA chains, by hydrophobic organic carbon enriched tail group interactions. Above described surface modification was intended to encourage fairly irreversibly bound octadecylammonium onto zeolite. To attain equilibrium about 30 hr time period was necessary for oxyanions adsorption onto inorganically exchanged zeolite, whereas about 3 days for organomodified zeolite.

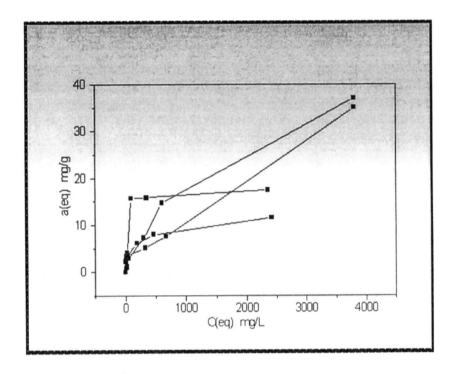

Figure 6. Adsorption isotherms for arsenate aqueous solutions with Ag- and Pb-clinoptilolites
and chromate aqueous solutions with Ag- and Pb-clinoptilolites, Tconst.= 23°C   (downwards
from the right)

Some differences in arsenate and chromate adsorption on ODA-
clinoptilolite and Pb-(Ag-linoptilolites) as well were recorded (Figs. 5 and
6). ODA-clinoptilolite exhibited more efficient arsenate and chromate
removal from aqueous solutions than the inorganically exchanged
modifications. However, silver exchanged clinoptilolite revealed higher
capacity values for both oxyanions uptake than lead exchanged clinoptilolite
did. This phenomenon supports preferred silver treated clinoptilolite
utilization for specific water purification process even on the base of
environmental acceptability.

The rate of adsorption from dilute aqueous solutions by solid adsorbents
(zeolites) is a highly significant factor for applications of this process for
water quality control.

Generally it can be stated that in rapidly stirred, batch type systems the
rate of uptake is controlled primarily by the rate at which adsorbate (solute)
is transported from the exterior to the interior sites of the zeolite particles.

The amount of solute adsorbed per unit weight of solid adsorbent, as a function of the concentration of solute remaining in solution at equilibrium and constant temperature is termed an adsorption isotherm.

Predominantly, Freundlich's fitted adsorption isotherms computed by means of simple linear regression were proposed for the mathematical description of the process studied. Unlike the Langmuir equation, the Freundlich model did not reduce to a linear adsorption expression at very low nor very high solute concentrations, as above resulted.

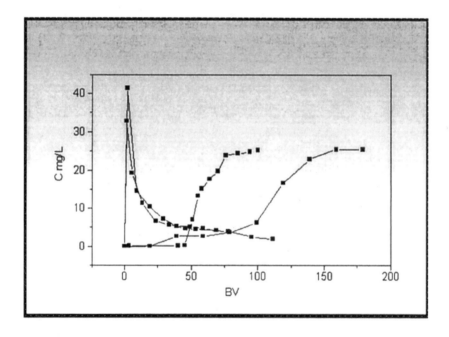

Figure 7. Regeneration of ODA-clinoptilolite columns loaded with chromate by means of 2% NaCl and 2% Na2SO4 aqueous solutions and breakthrough curves for ODA- clinoptilolite in 0.5 mM/L chromate solution by 30 BV/hr and 15 BV/hr in downflow mode (from the left)

By batch description trials Organo- and inorganically- modified zealot was subjected up to 24 hr in distilled water, tap water and 2% Nalco aqueous solutions in laboratory shaken machine to demonstrate how strongly the examined oxyanions are bound on the modified zeolite. While only slightly chromate desorption in the maximum extent about 20 mg/L was observed, approximately one order higher arsenate desorption was found, corresponding to increased ionic strength in waters. However, in both cases ODA-clinoptilolite exhibited the lowest desorption characteristics. Here, the desorbed anion concentrations negligible differed for above proposed elution reagents.

Fig. 7 presents partial results of dynamic regime experiments for chromate adsorption and desorption by ODA-clinoptilolite. As shown by breakthrough curves, ODA-clinoptilolite column quantitatively removes chromate species from "simulated waste water", apparently more efficiently by lower flow rate. Consequently to similar configuration of chromate and sulfate molecules, such loaded column was more efficient to regenerate with Na_2SO_4 than NaCl solution, as elution curves at the Fig. 7 illustrate.

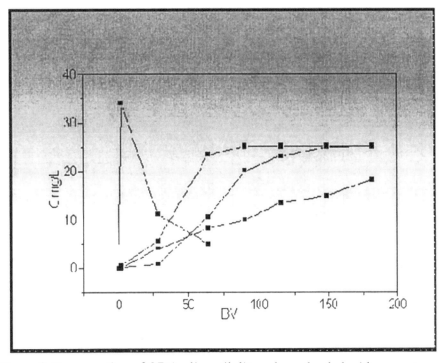

Figure 8. Regeneration of ODA-clinoptilolite column loaded with arsenate by means of 2% NaCl aqueous solution and breakthrough curves for ODA-clinoptilolite in arsenate solution of co = 25 mg/L; repeated cycle after regeneration, first cycle; breakthrough curve on Pb-clinoptilolite (from the left).

Arsenate removal by ODA-clinoptilolite proceeded almost analogous as chromate removal did, however the front part of the breakthrough curve is fairly shallow and indicates earlier leakage of pollutant into adsorbate (Fig.

8). Nevertheless, the reproducible repeated breakthrough curve was recorded after the exhausted column regeneration. Inorganically treated clinoptilolite (e.g. Pb-exchanged) did not prove such characteristic breakthrough profile by dynamic arsenate uptake as the organoclino-ptilolite did. Finally, Fig. 9 presents competitive anions $(SO_4^{2-}>Cl^->NO_3^-)$ inffluence on the chromate uptake by ODA-clinoptilolite. As showed, 1g sulfate and all examined anions together addition into current chromate solution exhibit almost the same breakthrough curves, what means that chloride and nitrate compete for available surface sites only negligible.

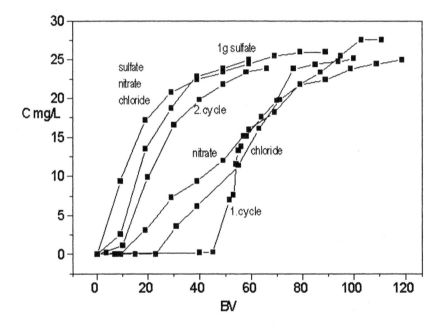

Figure 9. Breakthrough curves for chromate on ODA-clinoptilolite with competitive anions

4.   CONCLUSION

Some approach to tailor a new natural anion exchanger from ODA-modified clinoptilolite was studied. We verified repeatedly clinoptilolite enhanced oxyanion uptake mechanism after regeneration and so far confirmed the feasibility at laboratory level. To set this process for water treatment in recycling operation, a pilot plant research is necessary to accomplish. In contrast to clay's properties, which surface have been usually altered by surfactant modifiers, similarly modified natural zeolite represent fairly large potential for environmental applications especially on the base of superior hydraulic characteristics.

ACKNOWLEDGMENT

We acknowledge a partial support of the research by the National Science Council GAV under Contract No. 1/8049/01 and 1/8312/01.

LIST OF SYMBOLS AND ABBREVIATIONS

a	adsorption capacity in mg/g
a_{max}	maximum adsorption capacity
A, B	regression coefficients
b, K, n	isotherm constants
c_{eq}	equilibrium solute concentration in mg/L
ODA	octadecylammonium
r	correlation coefficient
SEM	scanning electron micrograph

REFERENCES

1. Förstner U., Wittmann G.T.V., *Metal Pollution in the Aquatic Environment,* Springer Verlag, Berlin – Heidelberg – New York, 1979, 3-40.
2. Phillips D. J.H., Rainbow P.S., *Biomonitoring of Trace Aquatic Contaminants,* Elsevier Applied Science, London and New York, 1993, 25-33.
3. Nriagu J.O., Davidson C.I., *Toxic Metals in the Atmosphere,* John Wiley & Sons, New York, 1986, 10-40.
4. Chmielewská-Horváthová E., Lesný J., *Geologica Carpathica-Series Clays 1,* 1992, 47-50.
5. http://www.who.int/inf-fs/en/fact210.html; Arsenic in drinking water, Fact Sheet No. 210, May 2001.
6. Henderson P., *Inorganic Geochemistry,* Pergamon Press, Oxford, 1982, 360.
7. Jesenák K., *Acta F.R.N. Chimia 39,* 1991, 45- 50.
8. Kozáč J., Očenáš D., *Mineralia slovaca* 14(1982) 549-552.
9. Horváthová E., *Environment Protec. Eng.* 16, 1990, 93-102.
10. Haggerty G. M., Bowman R.S., *Environmental Science & Technology* 28, 1994, 452-458.
11. Pfeiffer H., *Surface Phenomena Investigated by Nuclear Magnetic Resonance,* 1976, Physics Reports, North-Holland Publishing Company.

TESTING OF SORPTION MATERIALS FOR ARSENIC REMOVAL FROM WATERS

Birgit Daus, Holger Weiss
UFZ-Umweltforschungszentrum Leipzig-Halle GmbH, Interdisciplinary Department Industrial and Mining Landscapes, Permoserstrasse 15, 04318 Leipzig, Germany, E-mail: daus@pro.ufz.de

Abstract: Different materials (granulated activated carbon, Zr loaded activated carbon, Absorptionsmittel 3, zero-valent iron, granular iron hydroxide) were investigated for their capability to remove arsenic from water. Both arsenite and arsenate were investigated and batch and column tests were carried out.

Key words: Arsenic, sorption, water cleaning

1.   INTRODUCTION

Arsenic is a common contaminant in mine waters, in seepage waters from mine tailings and common in surface and groundwater surrounding such sites. Due to the high toxicity, especially of arsenite, the drinking water regulations have low thresholds for this element (e.g. Germany and USA: 0.01 mg/l).

Effective sorption materials for arsenic are needed for decentralized water cleaning or for effluent treatment and downstream remediation (on site or *in situ*) of contamination sources.

The background of this study is the investigation at the site Bielatal (see Daus et al. this volume). The arsenic concentrations in the seepage water of this tailings pond are high (up to 4 mg/l) and natural arsenic precipitation processes are incomplete. The neutral pH, the presence of both arsenite and arsenate (sum 1 mg/L), and the oxygen saturation of the water are the boundary conditions of the described experiments.

2.   MATERIALS TESTED

Five material were chosen for the following tests to include different sorption mechanisms as well as different active phases.
- Granular activated carbon (AC)

This material was chosen for comparison purpose for the following Zr-loaded activated carbon. The poor sorption capacity of this kind of material was shown before [1]. A carbon of the company CHEMVIRON CARBON (FILTRASORB TL 830) was used.
- Zr-loaded activated carbon (Zr-AC)

This sorption material was suggested by Peräniemi et al. [2] for analytical enrichments of phosphate and arsenate. The activated carbon was shaken with a solution of $ZrO(NO_3)_2$ as described in [2] and a final concentration of 28 mg Zr per g activated carbon was yield.
- Absorptionsmittel 3 (AM3)

This porous material is a commercial sorption material of the company Dr. Ecker GmbH (GERMANY) which is a composition of calcite, brucite, fluorite and iron hydroxides (RFA: 14 % Fe_2O_3).
- Zero valent iron (Fe0)

The granular cast iron of the company ERVIN AMSTEEL was chosen to include another type of material, having been described for arsenic removal from water [3,4] before.
- Granular iron hydroxide (GIH)

This special form of granular iron hydroxide consists of sand grains coated with iron hydroxides. This material has very good mechanical properties and a high content of the active phase iron hydroxide (RFA: 56 % Fe_2O_3).

3.   EXPERIMENTS

3.1   Batch Experiments

An amount of 0.5 g of sorption material was shaken with 50 mL of carbonate buffered water (pH = 7, 10 mg/L sulfate, 14 mg/L chloride) containing 500 µg/L of each As(III) and As(V). Samples were taken in increasing time intervals to investigate the kinetics of the sorption process. The arsenic species were measured by IC-ICP-MS [5].

3.2   Column Experiments

Columns with volume of 125 m³ were filled with different materials and the same water as described above was continuously percolated through the material. The flow rate was about 2 - 3 mL/min. Inlet and outlet water was sampled and analyzed for arsenic species daily.

4.   RESULTS AND DISCUSSIONS

For a first assessment of the performance of the different materials, batch experiments were carried out. The kinetics of the sorption processes of arsenic onto the different materials should give an indication of their efficiency. Figure 1 shows the results for the measured As(V) concentrations in dependence on time. The activated carbon gives poor results, as expected. However, the Zr loaded activated carbon shows a rapid reaction. The zirconyl ions at the surface of the activated carbon are a highly efficient phase for the sorption of arsenate. The half-life of this sorption reaction was < 10 min.

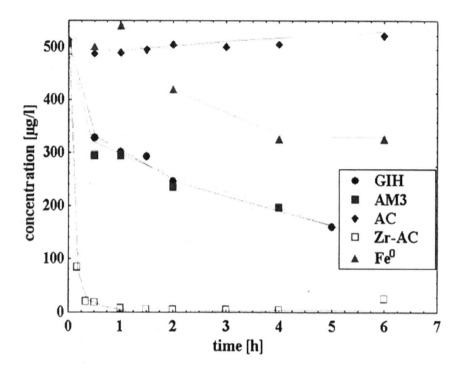

Figure 1. Arsenate sorption kinetics, batch experiments (experimental details see text)

The decrease of the arsenate concentration in presence of the Fe^0 is delayed, caused by a slow corrosion. The forming of iron hydroxide (active phase) at the surface of the particles is a requirement for the bond of arsenic to this material under oxidizing conditions. The other materials, containing already active iron phases (Absorptionmsmittel 3 and granular iron hydroxide), have this advantage over the Fe^0. Both materials show similar kinetics. After 24 hours the arsenate concentrations are lower than 50 µg/l in both cases. The removal of arsenite by the different materials is slow and seems to depend on the slow oxidation rate of arsenite to arsenate.

Column tests were carried out to investigate the behavior of the materials in a dynamic system. Figure 2 shows the results of these experiments for granular iron hydroxide, Absorptionsmittel 3 and Zr loaded activated carbon. The results of Fe^0 and activated carbon are not shown in the graph because they had initial high outlet concentrations.

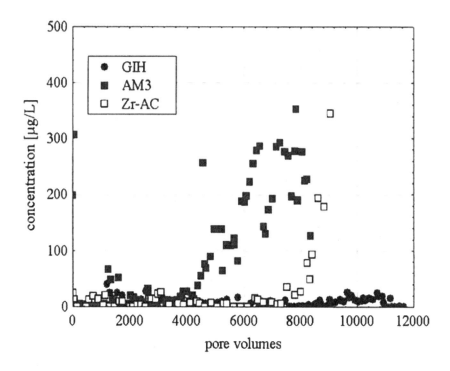

Figure 2. Column experiments, arsenate concentrations in the outlet of the columns

A typical breakthrough curve was observed for the column filled with Zr-loaded activated carbon after about 8000 pore volumes. This correspond to a uptake of 2.8 mg As/g. The concentrations in the outlet of the column with Absorptionsmittel 3 increased after about 4000 pore volumes, but no typical breakthrough curve was observed. The uptake until this point was only 2 mg As/g that is much lower than it was determined in the batch experiments. An explanation for this early increasing of the concentrations may be the high flow rate in comparison of the slow kinetics. The best results gave the column filled with the granular iron hydroxide. No breakthrough was observed up to now (12,000 pore volumes) and an uptake of about 2 mg As/g could be measured. The arsenite concentrations in the outlet of all three columns were very low and indicate an oxidation reaction.

Consequently, the granular iron hydroxide seems to be suitable for an on site or *in situ* treatment of arsenic contaminated surface waters. However, the experiment is not yet finished and some more experiments (influence of the dissolved iron, regeneration of the material, and surface characterization) are required before the material can prove its efficiency in a pilot test in the filed.

REFERENCES

1. Huang C.P., Fu P.L.K., Treatment of arsenic(V)-containing water by the activated carbon process, *J. Water Pollut. Control Fed.* 56, 1984, 233-242.
2. Peräniemi S., Hannonen S., Mustalathi H., Ahlgrén M., Zirconium-loaded activated charcoal as an adsorbent for arsenic, selenium and mercury, *Fresenius J. Anal. Chem.* 349, 1994, 510-515.
3. Lackovic J.A., Nikolaidis N. P., Dobbs G. M., Inorganic Arsenic Removal by Zero-Valent Iron, *Environmental Engeneering Science* 17, 2000, 29–39.
4. Su C., Puls R.W., Arsenate and arsenite removal by zero-valent iron: kinetics, redox transformation, and implications for in situ groundwater remediation, *Environ. Sci. & Technol.* 35, 2001, 1487-1992.
5. J. Mattusch and R. Wennrich (1998): Determination of anionic, neutral, and cationic species of arsenic by ion chromatography with ICPMS Detection in environmental samples.- Anal. Chem. 70, 3649 - 3655.

ZIRCONIUM(IV) LOADED DIAION CRP200 RESIN AS A SPECIFIC ADSORBENT TO AS(III) AND AS(V)

Akinori Jyo, Shuko Kudo, Xiaoping Zhu, and Kazunori Yamabe
Department of Applied Chemistry and Biochemistry, Faculty of Engineering, Kumamoto University, Kumamoto 860-8555, Japan, E-mail: jyo@gpo.kumamoto-u.ac.jp

Abstract: Performances of Zr(IV) loaded Diaion CRP200 (crosslinked polystyrene based methylenephosphonic acid resin) as specific adsorbent for phosphate, arsenate, and arsenite were evaluated by columnar approach. Phosphate and arsenate were adsorbed under weakly acidic conditions. Adsorbed phosphate and arsenate were quantitatively eluted by 0.1 M aqueous sodium hydroxide. After the elution operation, the adsorbent was able to regenerate by supplying a dilute sulfuric acid solution containing Zr(IV) to the column. Interestingly, breakthrough capacities were enhanced by repeating these adsorption-elution-regeneration operations up to 10 times and attained to the stationary ones of ca. 0.14 – 0.15 mmol/ml-Resin at until 1% breakthrough points. In addition, the proposed adsorbent was able to take up selectively both phosphate and arsenate even in the presence and chloride and sulfate in large excess. The most outstanding performances were observed in the adsorption of arsenite after complete washing the alkali-treated column with water; arsenite was much more strongly adsorbed than phosphate and arsenate. Then, the regeneration of the column was able to achieve by complete washing with water. The adsorption-elution-regeneration cycle was totally repeated by more than 40 times for ca. 2 years but no deterioration of the column performances was observed.

1.   INTRODUCTION

In Ganges Delta, groundwater contaminated with arsenic causes health risk of 40 million people.[1] Since arsenic exists as both arsenite (As(III)) and arsenate (As(V)), removal methods for both arsenic species are desired to resolve the arsenic calamity in Ganges Delta.[2] Granular adsorbents specific to arsenite and arsenate seem to be most promising because of high capacities and selectivity. Many polymeric adsorbents specific to both arsenic species have been proposed on the basis of ligand exchange approaches. However, most works focused on their behavior in batchwise adsorption of both arsenic species, and unfortunately their column-mode performances has not been widely studied.[3-13]

In a previous paper,[14] we have reported behavior of Zr(IV) loaded phosphonic acid resin (RGP) in the column-mode adsorption of arsenate and arsenite by repeating the adsorption-elution-regeneration cycles. The proposed Zr(IV) loaded RGP ligand exchanger gave promising results in the long term column-mode removal of arsenate and arsenate.[14] However, the resin RGP is not commercially available yet. However, we have found that Zr(IV) is also strongly retained by methylenephosphonic acid resins,[15] and then Zr(IV) loaded phosphonic acid resins similarly functions as Zr(IV) loaded RGP. In Japan, a crosslinked polystyrene based methylene-phosphonic acid resin called Diaion CRP200 (Scheme 1) is now commercially available. Then, the present work was planned to clarify properties of Zr(IV) loaded Diaion CRP200 as adsorbent for both arsenate and arsenite. The column packed with Zr(IV) loaded Diaion CRP200 was tested for the removal of phosphate, arsenate, and arsenite in successive without change of the packed resin. The column performances were evaluated by repeating the adsorption-elution-regeneration cycles by more than 50 times for about 20 months from April 2000 to December 2001.

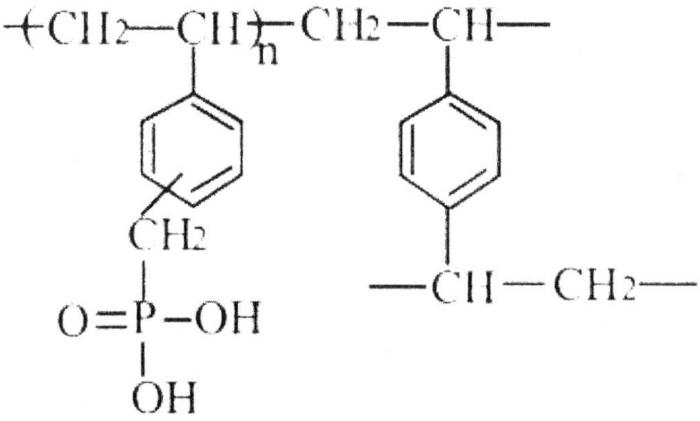

Scheme 1. Structure of Diaion CRP200

2.   EXPERIMENTAL

2.1   Materials

A sample of phosphonic acid resin Diaion CRP200 in wet Na^+ form was kindly provided from Mitsubishi Chemical Co. Ltd. This wet resin sample was dried in vacuum, and then the resin with particle sizes of 60 – 32 mesh was selected by meshing. The selected resin was conditioned by treatment with 1 M HCl, water, 2 M NaOH, water, 1 M HCl, and water in successive. Finally, thus, the resin was changed into H^+ form. The phosphorus content and acid capacity of the selected resin in the H^+ form were measured according to the reported methods and were 4.6 mmol/g and 8.8 meq/g, respectively. Wet volume of the resin was 2.6 ml/g. Hereafter, this is abbreviated as CRP200 for simplicity.

All reagents used were of guaranteed grade and ultra pure water was used, unless otherwise noted. Seawater sample was provided by the Aitsu Marine Biological Station of Kumamoto University (Amakusa, Kumamoto, Japan). It was filtered with a Millipore filter (0.45 m) to remove suspending particles.

2.2   Preparation of Zr(IV) loaded Diaion CRP200 column

A wet settled CRP200 in the H^+ form (1 ml) was packed into a glass column (i.d. 0.7 cm). Then, H_2SO_4 (0.5 M) containing $Zr(SO_4)_2$ at 0.01 M (200 ml) was fed to the column, and then 0.5 M H_2SO_4 (20 ml) and water (20 ml) at a flow rate of 10 ml/h. All column effluents including the washings were collected, and amounts of Zr(IV) in the all column effluents were determined by means of ICP-AES analysis. By subtracting the amount of Zr(IV) in the all column effluent from that in the solution loaded, the amount of loaded Zr(IV) was determined to be 0.60 mmol/ml of wet resin. This column was used without change of the resin throughout.

2.3   Column operation for adsorption and elution of phosphate species

Since properties of both arsenate and phosphate species are very close, the adsorption of phosphate species was studied prior to that of arsenate. The phosphate adsorption was tested by the following two methods (i) and (ii). Hereafter, flow rate and volume of feed are given in space velocity (SV, h^{-1})

and bed volumes (BV), which are designated by ratios of flow rate in ml/h and volume of the feed (ml) to the wet resin volume in the column (1mL), respectively. Feeding solutions used in the adsorption operation were prepared by dissolving $Na_2HPO_42H_2O$ into water and the pH was adjusted with HNO_3.

2.3.1   Method (i):

The adsorption operation was conducted by feeding 1 mM phosphate solutions to the conditioned column at SV 10 or 20 h^{-1}, and then 20 BV of water at SV 3 h^{-1}. The elution operation consisted of feeding 0.1 M NaOH solution (80 BV) and then 20 BV of water at a flow rate of SV 3 h^{-1}. The regeneration of the column was conducted by feeding 0.5 M sulfuric acid (20 BV) and 20 BV of water at SV 10 h^{-1}. All column effluents including washings in the adsorption and elution operations were collected on a fraction collector, and concentrations of phosphorus and zirconium in each fraction were determined by ICP-AES. Volume of each fraction was 5 BV for the adsorption operation and 4 BV for the elution operation. However, column effluents in regeneration operations were not analyzed.

2.3.2   Method (ii):

The only difference from the method (i) was in the regeneration operation. Here, the regeneration operation was conducted by supplying 0.5 M sulfuric acid (5 BV), 0.5 M sulfuric acid containing zirconium sulfate at 0.01 M (10 BV), 0.5 M sulfuric acid (15 BV) and water (20 BV) in successive at SV 10 h^{-1}. The adsorption and elution operations were almost the same as those in the method (i).

2.4   Column operation for adsorption and elution of arsenate and arsenite

All procedures for removal of the arsenate were conducted by the method (ii). Feeding solutions used in the adsorption operation for arsenate were prepared by dissolving $Na_2HAsO_47H_2O$ into water, and those used in the arsenite adsorption were prepared by dissolving $NaAsO_2$ into water. The pH of these solutions was adjusted with HNO_3.

In the case of arsenite removal, on the other hand, the regeneration operation was quite different from that for the removal of phosphate or arsenate. After the column was treated with 0.1 NaOH (80 BV), the column was washed thoroughly with water until the pH of washing became nearly neutral. Then, the adsorption operation was started by feeding aqueous solutions of arsenite at SV 20 h^{-1} and then 20 BV of water at SV 3 h^{-1}. The elution of the adsorbed arsenite was conducted by supplying 1.5 M NaOH (ca. 100 BV). After the elution operation, the column was washed with water for the next adsorption operation of arsenite.

3.   RESULTS AND DISCUSSION

3.1   Behavior of Zr(IV)-loaded CRP200 in adsorption of phosphate species

Since properties of both phosphate and arsenate are very similar each other, the adsorption of phosphate was examined prior to the adsorption of arsenic species. Here, the feeding solution in the adsorption operation was 1 mM phosphate solution of pH3. Table 1 summarizes detailed experimental conditions and column performances during repeated adsorption-elution-regeneration cycles. Since supplied volumes of the feed are not constant (101 – 193 BV), it is not easy to judge the efficiency of the adsorption from total uptake of phosphate. Thus, removal of phosphate until 100 BV is listed at the last column of Table 1 as an index of the column performances.

Run	Feed Volume supplied BV[a]	Feed Concn of phosphate mM	pH	Flow rate h[-1]	Regeneration method of column	Total uptake of phosphate mmol/ml-R	Amount of Eluted phosphate mmol/ml-R	Recovery of eluted phosphate %	Removal of phosphate until 100 BV %
1-1	160	0.998	3.00	10	–	0.152(95.0)[b]	0.159	105	98.3
1-2	161	1.01	2.91	10	–	0.132(81.0)	0.132	100	92.7
1-3	152	1.00	2.97	10	–	0.117(77.0)	0.117	100	89.1
2-1	105	0.983	3.01	10	ii	0.0977(96.8)	0.106	106	96.9
2-2	101	1.02	2.98	10	ii	0.103(100)	0.110	107	100
2-3	115	1.01	3.01	10	ii	0.115(99.1)	0.108	93.9	100
2-4	146	1.01	3.01	10	ii	0.147(100)	0.141	95.7	100
3-1	193	1.00	2.98	20	ii	0.188(97.4)	0.185	98.9	99.7
3-2	177	1.02	3.05	20	ii	0.177(97.8)	0.171	95.6	100
3-3	163	1.02	2.96	20	ii	0.164(98.8)	0.163	99.3	100
3-4	172	1.01	2.96	20	ii	0.169(97.1)	0.168	99.4	99.0
3-5	143	0.991	3.02	20	ii	0.141(99.3)	0.136	96.3	99.9
3-6	150	0.996	3.03	20	ii	0.149(100)	0.151	102	99.7
3-7	157	0.982	2.99	20	ii	0.153(99.4)	0.154	100	99.7
4-1	162	0.906	3.05	20	ii	0.146(99.3)	0.164	112	99.9

a) Bed volumes b) Percents of adsorbed phosphate

Table 1. Conditions and results of adsorption-elution-regeneration cycles for removal of phosphate.

Results from runs from 1-1 to 1-3 are obtained by the method (i), in which the column was regenerated with 0.5 M H_2SO_4 without Zr(IV) after elution of the adsorbed phosphate with 0.1 M NaOH.. Clearly, the removal of phosphate until 100 BV markedly decreased with repeating the adsorption-elution-regeneration cycle. Since it was identified that the total loss of Zr(IV) in the adsorption and elution operations was less than 0.1 % of the initially loaded amount of Zr(IV), the decrease in uptake of phosphate may be ascribable to the loss of loaded Zr(IV) in the regeneration step; in which excess 0.5 M H_2SO_4 was supplied to the column for neutralization of hydroxide ions bound to loaded Zr(IV) during the elution operation. Then, the further studies were conducted by the method (ii), in which 0.5 M H_2SO_4 containing 0.01 M $Zr(SO_4)_2$ was used to regenerate the column to depress the dissolution of Zr(IV) and/or to load more Zr(IV) onto the column.

In runs from 2-1 to 2-4, the adsorption of phosphate was carried out by supplying the feeding solutions at SV 10 h^{-1}. Leakage of phosphate from the column up to 100 BV of the feed is markedly depressed by repeating the adsorption-elution-regeneration cycle of the method (ii). In the runs from 2-2 to 2-4, the uptake until 100 BV of the feed is almost quantitative, and then the flow rate of the feed increased to SV 20 h^{-1}. Even at the higher flow rate of SV 20 h^{-1} (runs from 3-1 to 3-7), the column took up phosphate almost quantitatively up to 100 bed volumes of the feed.

Table 1 also gives total uptake of phosphate and its amount eluted with 80 BV of 0.1 M NaOH. The recovery is almost quantitative. After the run 3-5, more than 99 % of the supplied phosphate were adsorbed even at high flow rate of SV 20 h^{-1} indicating that the column attained to the stationary state. Since phosphate concentration in the column effluent very gradually increased, the determination of breakthrough points with high precision was difficult to achieve. However, it can be said that even 1% breakthrough points are greater than 100 bed volumes after the run 3-5.

The final run 4-1 was the adsorption of phosphate from seawater to which 1 mM of phosphate was spiked. The concentration of NaCl in seawater is ca. 0.5 M, which is 500 times higher than the concentration of the spiked phosphate. Thus, it can be concluded electrolytes in seawater do not interfere with the adsorption of phosphate by Zr(IV) loaded CRP200 as in the case of arsenate uptake by Zr(IV) loaded phosphoric acid resin RGP.[14]

3.2   Removal of arsenate

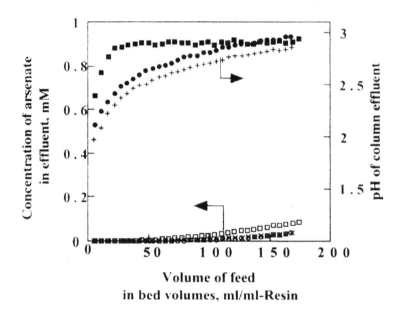

Figure 1. Breakthrough curves of arsenate and pH profiles of column effluents during adsorption from feeds in t he absence and presence of foreign anions.

Feed: 1 mM arsenate of pH 3. Flow rate 20 h-1 in SV.

Without added salt (run 1-3 in Table 2) ▢ As, ▢ pH. In the presence of 100 mM sodium chloride (run 2-3 in Table 2) ▢ As, ▢ pH. In the presence of 100 mM sodium sulfate (run 3-3 in Table 2) ▢ As, ▢ pH.

Since the amount of Zr(IV) on CRP200 was attained to plateau judged from the results mentioned in the preceding section, breakthrough behavior of arsenate was atated by supplying 1 mM arsenate solutions of pH 3 in the absence and presence of an electrolyte NaCl or Na$_2$SO$_4$.   Figure 1 illustratively shows breakthrough curves of arsenate and pH profiles of column effluents. Sodium sulfate slightly interferes with the adsorption of arsenate but not sodium chloride. The pH of the column effluents increases with an increase of the supplied volume of the feed and is finally asymptotic to pH 3, which is the pH of the feed. These pH profiles may means that hydrogen ion attached to sites unoccupied by Zr(IV) may be eluted with sodium ion in the feed. Figure 2 shows a typical elution curve of arsenate with 0.1 M NaOH, suggesting that the quantitative elution is achieved by 60

BV of the eluent. Table 2 summarizes detailed experimental conditions and column performances. Since volumes of the feed supplied to the column were from 145 to 174 BV, percentages of removed arsenate until 140 BV of the feed is listed. These data means that interference by both anions was not serious up to 140 BV of the feed..

run	Feed				Total uptake of arsenate mmol/ml-R[b]	Amount of arsenate eluted mmol/ml-R	Recovery of eluted arsenate %	Removal of arsenate until 140 BV %
	Volume supplied BV[a]	Concn of arsenate mM	pH	Electrolyte added mM				
1-1	145	1.02	3.08	none	0.147(99.3)[b]	0.131	89.1	99.4
1-2	154	1.01	3.10	none	0.154(98.7)	0.151	98.2	99.0
1-3	167	1.02	3.11	none	0.168(98.8)	0.166	98.9	99.4
2-1	164	0.998	3.07	NaCl. 100	0.162(98.8)	0.152	93.9	99.5
2-2	166	0.958	3.06	NaCl. 100	0.158(99.3)	0.161	102	99.9
2-3	169	0.982	3.04	NaCl. 100	0.162(97.6)	0.160	99.0	99.7
3-1	148	0.998	3.11	Na2SO4. 100	0.146(98.6)	0.144	98.3	99.5
3-2	162	0.967	3.11	Na2SO4. 100	0.153(97.5)	0.160	104	99.0
3-3	174	0.991	2.94	Na2SO4. 100	0.167(97.1)	0.165	99.1	98.0

[a] Bed volumes　[b] Percents of adsorbed arsenate

Table 2. Conditions and results of adsorption-elution-regeneration cycles for removal of arsenate in the absence and presence of sodium chloride or sodium sulfate.

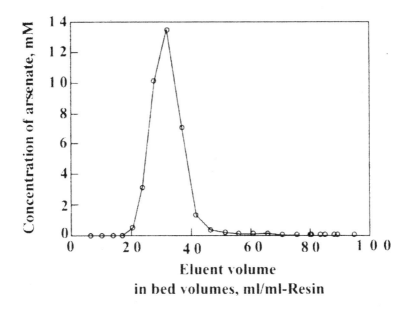

Figure 2. Typical elution curve of arsenate. Eluent 0.1 M sodioum hydroxide, flow rate SV 3 h-1. Elution in run 1-3 in Table 2.

Table 3 summarizes the results for the effect of pH of the feeds on uptake of arsenate. Uptake or removal of arsenate until 140 BV of the feed shows that the uptake slightly decreases with an increase of the pH of feeds as widely observed in ligand exchange uptake of arsenate by metal and/or metal oxide loaded ligand exchangers. [3-15]

Run	Feed			Total uptake of arsenate	Amount of eluted arsenate	Recovery of eluted arsenate	Removal of arsenate until 140 BV
	Volume supplied (BVa)	Concn of arsenate mM	pH	mmol/ml-R	mmol/ml-R	%	%
1-1	281	1.03	4.06	0.259(89.6)b)	0.251	97.0	98.5
1-2	156	1.04	4.05	0.161(99.4)	0.157	97.6	98.9
2-1	161	0.998	5.04	0.156(97.0)	0.154	98.9	98.8
2-2	168	1.05	4.95	0.171(97.2)	0.166	97.3	98.3
3-1	178	0.980	5.93	0.162(93.1)	0.177	109	96.2
3-2	146	0.998	5.80	0.144(98.6)	0.146	101	98.8
3-3	162	0.960	5.85	0.151(96.8)	0.159	105	97.4

a) Bed volumes b) Percents of adsorbed arsenate

Table 3. Effect of pH of the feed on the removal of arsenate.

3.3   Removal of arsenite

We have reported that Zr(IV) loaded phosphoric acid RGP exhibits high selectivity to arsenite when it is washed with water thoroughly after its alkaline treatment.[14] The quite similar phenomenon was observed in the case of Zr(IV) loaded CRP200 as shown in Fig. 3. The pH of the effluent from incompletely washed column is nearly equal to or higher than 10, resulting in the incomplete removal of arsenite. On the other hand, the pH of the effluent from the well washed column was less than 9 and nearly equal to neutral pH. Clearly, arsenite was much more effectively removed by the well washed column than the incompletely washed case.

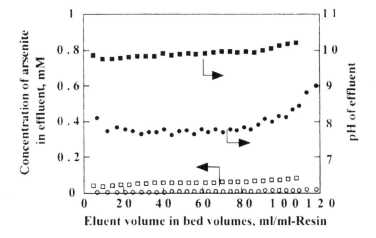

Figure 3 Effect of washing of alkali-treated column on the adsorption of arsenite.
Completely washed column: ▢ arsenite, ▢ pH (run 1-1 in Table 4).
Incompletely washed column: ▢ arsenite, ▢ pH(run 1-5 in Table 4).

Run	Feed			pH range of Column effluent	Total uptake of arsenite mmol/ml-R	Amount of eluted arsenite mmol/ml-R	Recovery of eluted arsenite %	Removal of arsenite until 70 BV %
	Volume supplied BV[a]	Concn of arsenite mM	pH					
1-1	108	1.04	8.68	9.8 - 10.2	0.106(94.6)[b]	0.0460	43.4	95.2
1-2	89	0.992	7.33	6.9 - 7.4	0.0883(100)	0.0987	112	99.7
1-3	134	1.01	8.51	8.0 - 9.6	0.134(99.3)	0.143	107	99.2
1-4	72	1.03	8.00	7.4 - 7.5	0.0737(99.3)	0.0639	86.7	99.7
1-5	118	0.984	7.92	7.7 - 9.0	0.115(99.1)	0.121	105	99.3

[a] Bed volume   [b] Percents of adsorbed arsenite

Table 4. Conditions and results of repeated removal of arsenite.

Table 4 summarizes experimental conditions and results of 5 cycles of the adsorption-elution operation. Different from the case of arsenate, the regeneration procedure by sulfuric acid solution of Zr(IV) was not required, since leak of Zr(IV) from the column becomes to be negligibly small above pH 3. The run 1-1 in Table 4 gave the low recovery of the eluted arsenite (43.4 %). This means that 80 BV of the 0.7 M NaOH is not enough to quantitative elution of arsenite. Consequently, the elution was conducted by using ca. 100 BV of 1.5 M NaOH. Figure 4 shows a typical elution curve of arsenite. Prolonged tailing is observed indicating that arsenite binds much more strongly to Zr(IV) than does arsenate. Indeed, arsenate is much more easily eluted even with 0.1 M NaOH as shown in Fig. 2. Therefore, it is expected that the interfering effect of NaCl and Na_2SO_4 on the arsenite adsorption will be very minor. Table 5 summarizes experimental conditions and results for the effect of NaCl and Na_2SO_4 on the removal of arsenite. No interference from these salts was observed. On the contrary, the arsenite removal was somewhat enhanced by the salts. Thus, Zr(IV) loaded CRP200 exhibits extremely high selectivity to arsenite in the neutral pH region.

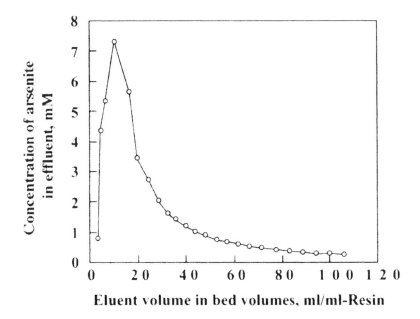

Figure 4. Typical elution curve of arsenite.
Eluent 1. 5 M sodium hydroxide, SV 3 h-1. The elution of run 1-1 in Table 5.

Run	Feed				Total uptake of arsenite mmol/ml-R	Amount of eluted arsenite mmol/ml-R	Recovery of eluted arsenite %	Removal of arsenite until 130 BV %
	Volume supplied BV[a]	Concn of arsenite mM	pH	Electrolyte added mM				
1-1	164	1.08	7.36	NaCl, 100	0.175(98.9)[b]	0.177	101	99.7
1-2	159	0.911	8.03	NaCl, 100	0.142(97.9)	0.163	114	98.8
2-1	150	0.986	7.76	Na$_2$SO$_4$, 100	0.147(99.3)	0.151	102	99.9
2-2	141	1.14	8.07	Na$_2$SO$_4$, 100	0.161(100)	0.159	98.9	99.9

[a] Bed volumes. [b] Percents of adsorbed arsenite.

Table 5. Effect of sodium chloride and sodium sulfate on the removal of arsenite.

3.4   Removal of dilute arsenite

In this work, the column performances were examined by supplying 1 mM of arsenic species, which corresponds to 75 ppm of arsenic. Such a high level arsenic is rarely found in surface water. Since the highest concentration levels of arsenic in well water of Ganges Delta are ca. 2 ppm. Thus, the removal of arsenic form dilute arsenite solutions was tested. Figure 5 illustratively shows the results.

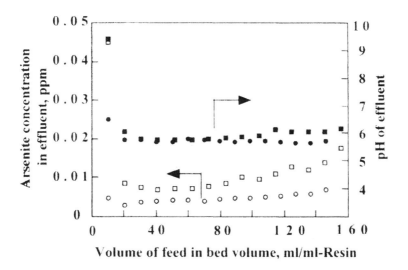

Figure 5. . Removal of arsenite from dilute solution.
Feed: 2.24 ppm As (arsenite), flow rate SV 20 h-1, pH 6.55, ☐ As, ☐ pH.
Feed: 2.55 ppm As (arsenite), flow rate SV 10 h-1, pH 6.48, ☐ As, ☐ pH.

When the feed was supplied to the column at the flow rate of SV 10, the concentration of arsenic in the column effluent was lower than 0.01 ppm, which is the level of arsenic in drinking water set by the World Health Organization. [1] The pH of the column effluent was also around 6 in the case of the flow rate of the feed was 10 h^{-1}. On the other hand, the concentration of arsenic in the column effluents tends to increase with an increase in flow rate of feeds as judged from Fig. 5. Then, flow rates higher than SV 20 h^{-1} seems to be inadequate.

As conclusion, Zr(IV) loaded Diaion CRP200 will be promising in the removal of arsenite and arsenate, whereas much more detailed studies will be needed about its behavior in the removal of sub-ppm levels of both arsenate and arsenite. However, we would like to emphasize that all results described here were obtained by using the single column during ca. 2 years

and more than 40 cycles of the adsorption-elution-regeneration operation.
No deterioration of the resin was observed.

REFERENCES

1. Karim, M. M., Arsenic in groundwater and health problems in Bangladesh, *Wat. Res.*, 34, 2000, 304-310.
2. Nickson, R. T., McArther, J. M., Ravenscroft, P., Burgess, W. G., Ahmed, K. M., Mechanism of arsenic release to groundwater, Bangladesh and Wet Bengal, *Appl. Geochem.*, 15, 2000, 403-413.
3. Yoshida, I., Konami, T., Shimonishi, Y., Morise, A., Ueno, K., Adsorption of arsenic (III) ion on various ion exchange resins loaded with iron (III) and zirconium (IV). *Nippon Kagaku Kaishi*, 1981, 379-384.
4. Chanda, M., O'Driscoll, K. F., Rempel, G. L., Ligand exchange sorption of arsenate and arsenite anions by chelating resins in ferric ion form; I. Weak-base chleating resin Dow XFS-4195. *Reactive Polym.* 7, 1988, 251-261.
4. Chanda, M., O'Driscoll, F. K, Rempel, G. L., Ligand exchange sorption of arsenate and arsenite anions by Chelating resins in ferric ion form; II. Iminodiacetic chelating resin Chelex 100. *Reactive Polym.* 8, 1988, *85.*
5. Maeda, S., Ohki, A., Saikoji, S., Naka, K., Iron(III) hydroxide-loaded coral lime stone as an adsorbent for arsenic(III) and arsenic(V), Sep. Sci. Technol., 27, 1992, 681- 689.
6. Matsunaga, H., Yokoyama, T., Eldridge, R. J., Bolto, B. A., Adsorption characteristics of arsenic(III) and arsenic(V) on iron(III)-loaded chelating resin having lysine- N^α, N^α-diacetic acid moiety. *Reactive & Functional Polym.* 29, 1996, 167-174.
7. Min, J. H., Hering, J. G., Arsenate sorption by Fe(III)-doped alginate gels., Wat. Res., 32, 1998, 1544-1552.
8. Haron, M. J., Wan Yunus, W. M. Z., Yong, N. L., Tokunaga, S., Sorption of arsenate and arsenite anions by iron(III)-polu(hydroxamic acid) complex, Chemoshpere, 39, 1999, 2459-2466.
9. Kobayashi, E., Sugai, M., Imajyo, S., Adsorption of arsenate ion by the zirconium(iv) oxide hydrate active carbon complex. *Nihon Kagaku Kaishi* 1984, 656-660.
10. Suzuki, T. M., Bomani, J. O., Matsunaga, H., Yokoyama T., Preparation of porous resin loaded with crystalline hydrous zirconium oxide and its application to the removal of arsenic. *React. Funct. Polym.* 2000, *43*, 165-172.
11. Suzuki, T. M., Tanaka, D. A. P., Tanco, M. A. L., Kanesato, M., H., Yokoyama, T., Adsorption and removal of oxo-anions of arsenic and selenium on zirconium(IV) loaded polymer resin functionalized with diethylenetriamine-N, N,N',N'-polyacetic acid, J. Environ. Monit., 2, 2000, 550-555.
12. Suzuki, T. M., Tanco, M. L., Tanaka, D. A. P., Yokoyama, T., Matsunaga, H., Yokoyama, T., Adsorption characteristic and removal of oxo-anions of arsenic and selenium on the porous polymers loaded with monoclinic hydrous zirconium oxide, Sep. Sci. Technol., 36, 2001, 103-111.
13. Dambies, L., Guibal, E., Rose, A., Arsenic (V) sorption on molybdate-impregnated chitosan beads, Colloids and Surfaces A, 170 (2000), 19-31.
14. Zhu, X., Jyo, A., Removal of arsenic(v) by zirconium(IV) loaded phosphoric acid chelating resin, *Sep. Sci. Technol.* , 36, 2001, 3175-3189.

15. Yamabe, K., Jyo, A., Development of specific adsorbent for fluoride based on metal ion loaded methylenephosphonic acid resin, in: *Proceedings of the 12 th International Conference on Chemistry for the Protection of the Environment,* eds. Z. Cao, and L. Pawlowski, Nanjing University Press, Nanjing, China, 1999, pp.230-237, ISBN 7-305-03345-6/O-233.

BIFUNCTIONAL CATION EXCHANGE FIBERS HAVING PHOSPHONIC AND SULFONIC ACID GROUPS

Akinori Jyo, [*1] Kenji Okada, [1] Masao Tamada, [2] Tamikazu Kume, [2] Takanobu Sugo[2], and Masato Tazaki[3]
*Akinori Jyo, [*1] Kenji Okada, [1] Masao Tamada, [2] Tamikazu Kume, [2] Takanobu Sugo[2], and Masato Tazaki[3]*

Abstract: Bifunctional cation exchange fibers were derived from polyethylene coated polypropylene fibers (PPPE). Two types of PPPEs were used. One is short cut fiber (PPPF-f, 0.9 denier) and the other non-woven cloth (PPPF-c, 1.5 denier). First, precursory fibers were prepared by graft copolymerization of chloromethylstyrene and styrene onto PPPE-c and PPPE-f by electron beam pre-irradiation induced liquid phase graft polymerization technique. Second, the precursory fibers were functionalized by Arbusov reaction, sulfonation, and acid hydrolysis in successive, resulting in the bifunctional fibers FPS-c (from PPPE-c) and FPS-f (from PPPE-f) having both phosphonic and sulfonic acid groups. For comparison, monofanctional phosphonic acid fibers FP-c (from PPPE-c) and FP-f (from PPPE-f) were also prepared by Arbusov reaction and the subsequent acid hydrolysis. Batchwise study using FPS-c and FP-c clarified that the bifunctional fiber FPS-c exhibits the characteristic metal ion selectivity resulting from the cooperative recognition of metal ions by both functional groups, and the bifunctional fiber takes up Pb(II) more rapidly than the monofunctional phosphonic acid fiber (FP-c) and resin (Diaion CRP200). Column-mode study using both bifunctional and monofunctional fibers FPS-f and FP-f revealed that both exhibit flow rate independent breakthrough profiles of Pb(II) up to flow rate of 900 h^{-1} in space velocity, indicating their extremely rapid adsorption rates of Pb(II). The bifunctional fiber gave breakthrough capacities of 0.54 – 0.57 mmol/g , which are nearly twice those of monofunctional one (0.28 – 0.33 mmol/g).

Key words: Ion exchange fiber, bifunctional fiber, sulfonic acid groups, phosphonic acid groups.

1.   INTRODUCTION

Phosphonic acid resins exhibit the highest selectivity to Pb(II) among common divalent metal ions and they also show the extremely high selectivity to metal ions classified into hard Lewis acid, such as Fe(III), Zr(IV), Mo(VI), and U(VI) .[1-4] Since these hard Lewis acid cations tend to form sparingly soluble hydroxides, they exist as soluble species in strongly acidic pH region. This means that the adsorption of these metal ions by phosphonic acid exchangers (resins and fibers) must be conducted in strongly acidic conditions. However, this brings shrinking of phosphonic acid resins, since polymer bound phosphonic acid groups are weak acid (pKa$_1$ is ca. 3). [5] This shrinking lowers both capacities and adsorption rates in uptake of metal ions. In order to resolve these problems, Alexandratos' group have proposed bifunctional phosphonic acid resins, which have sulfonic acid groups in addition to phosphonic acid ones.[6-10] Indeed, the bifunctional phosphonic acid resin Diphonix effectively adsorbs actinide ions and Fe(III) even from strongly acidic media.[7]

Recently, we[11, 12] have developed cation exchange fibers having phosphoric acid or phosphonic acid groups, which were derived from polyethylene coated polypropylene fibers having grafted poly(glycidyl methacrylate) or poly(chloromethylstyrene) chains. Although these fibers exhibited extremely rapid adsorption rates in the column-mode adsorption of divalent heavy metal ions, such as Pb(II) and Cu(II) from weakly acidic pH solutions (pH ca. 5 –6), [11-12] our preliminary works showed that these fibers lose their high performances in the adsorption of Mo(IV) and Zr(IV) from strongly acidic solutions below pH 2.[13]   Consequently, this work was planned to resolve these problems of fibrous phosphonic acid exchangers utilizing the Alexandratos's concept of bifunctional exchangers.[6-10] The bifunctional phosphonic acid fiber shown in Scheme 1 was prepared, and behavior of the resulting fibers in uptake of Pb(II) and Ba(II) was studied by means of both batchwise and columnar approaches. The metal ions Pb(II) and Ba(II) were selected, since polymer-supported sulfonic acid prefers Ba(II) to Pb(II) but polymer supported phosphonic acid show the reversed selectivity. [3,4] For the sake of comparison, monofunctional phosphonic acid fiber and resin were used, and structures of the exchangers used are also given in Scheme 1.

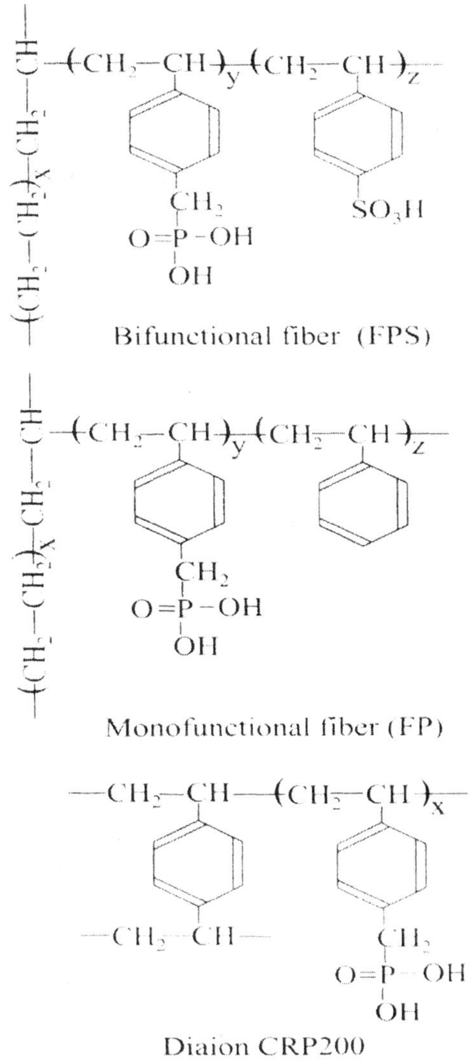

Bifunctional fiber (FPS)

Monofunctional fiber (FP)

Diaion CRP200

Scheme 1. Ion exchange fibers and resin used in this work.

2.   EXPERIMENTAL

2.1   Materials

Short cut polyethylene coated polypropylene fiber (0.9 denier, length 3.8 cm) and its non-woven cloth (1.5 denier) were provided by Kurasiki Textile

MFG Co., Osaka, Japan. Hereafter, the former and the latter are abbreviated as PPPE-f and PPPE-c, respectively. Chloromethylstyrene (CMS) was provided by Seimi Chemical Co., Chigasaki, Japan, and styrene (ST) was purchased from Wako Pure Chemical Co. Reagents were of guaranteed grade unless otherwise noted. Ultra-pure water was used throughout. A monofunctional phosphonic acid resin Diaion CRP200 was also used for the sake of the comparison.

2.2   Graft copolymerization of CMS and ST onto PPPE-c and PPPE-f

Procedures were almost the same as those for graft polymerization of CMS alone onto PPPE-c and PPPE-f.[11] The mixture of CMS (80 mol%) and ST(20 mol%) was diluted with dimethyl sulfoxide (50 wt% of the monomers). This was used as the bathing solution of electron irradiated PPPE-f and PPPE-f (2 MeV, 200 KGy) in graft copolymerization of both monomers. Electron irradiated PPPE-f and PPPE-f were immersed in the bathing solution for a given time at 40 °C. After the grafted PPPE-c and PPPE-f were washed with N,N-dimethylformamide and then with methanol, they were dried in vacuum. From an increase in weight (W_{in}) by graft polymerization, degree of grafting (dg) is designated as 100 (W_{in}/W_0). Here, W_0 is the weight of the trunk fibers before grafting. The chlorine contents of the grafted fibers were measured by decomposition with the Schöniger flask method, the product being adsorbed in hydrogen peroxide solution. The resulting chloride ion was determined by ion chromatography.

2.3   Functionalization

Scheme 2 shows preparation of the bifunctional fiber. PPPE-c and PPPE-f grafted with CMS and ST were functionalized by the reported method.[6] Here, an example of functionalization procedures is described. The precursory fibers (PPPE-f grafted with CMS and ST, 2 g) and triethyl phosphite (60 ml) were taken into a three-necked flask (200 ml) equipped with a Liebig condenser. After the mixture in the flask was heated for 48 h at 100 °C, the resulting phosphorylated PPPE-f was washed with acetone, aceton-water mixture and water, and dried in vacuum oven at 40 °C. In order to introduce sulfonic acid groups, the phosphorylated PPPE-f was treated with a 30 mL of 1,2-dichloroethane solution of chlorosulfonic acid (10 wt %) for 2 h at room temperature. Finally, diethyl phosphonate groups on the fibers were hydrolyzed with 12 M hydrochloric acid under refluxed conditions for 48 h. The functionalization of the grafted PPPE-c was also conducted by almost the same procedures. For the sake of comparison, monofunctional phosphonic acid fibers were prepared by acid hydrolysis of phosphorylated PPPE-c and PPPE-f. All functionalized fibers were washed with water and dried in a vacuum oven at 40 °C. Their phosphorus and sulfur contents and acid capacities were evaluated by reported methods.[14, 15] Introduction of each functional group was identified by measuring FTIR spectra. All ion exchangers (fibers and resin) were used in the hydrogen ion form unless other wise specified. Cloth type fibrous exchangers from PPPE-c were used in batchwise studies, and short cut fiber type ones from PPPE-f were used in column-mode study.

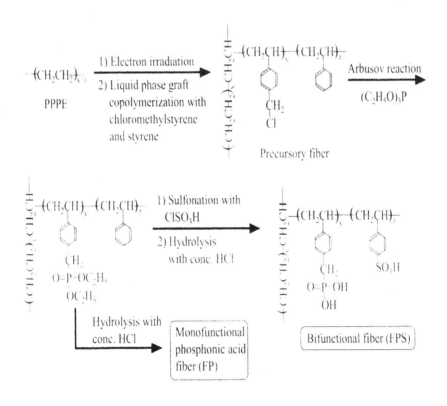

Scheme 2. Preparation of bifunctional fibers.

2.4   Batchwise study

2.4.1   Distribution ratios of metal ion.

An ion exchanger (0.04 g) and a metal ion solution (10^{-4} M, 25 ml) were taken into 50 ml Erlenmeyer flasks. Then, the flasks were shaken with a mechanical shaker at 30 °C for 24 h. The metal ion concentration in the aqueous phase was determined by means of ICP-AES. From a decrease of the metal ion concentration in the aqueous phase, uptake of metal ion in mmol/g was calculated. Here, only nitric acid was used in the pH adjustment.

2.4.2   Capacities for uptake of metal ions.

An ion exchanger (resin 0.05 g or fiber 0.1 g) and a metal ion solution (0.01M, 25 ml) were taken into 50 ml Erlenmeyer flasks. Subsequent procedures were almost the same as those mentioned above. In the pH adjustment, nitric acid (pH< 3), acetic acid and sodium acetate buffers (3 < pH < 5) were used.

2.4.3   Adsorption rate.

Into a three-necked flask (200 ml) equipped with a mechanical stirrer, an ion exchanger (0.3 g) and water (135 mL) were taken. The resulting mixture was allowed to stand overnight for swelling of the ion exchanger. Then, a lead nitrate solution (0.05 M, 15 ml) was added to the flask, and immediately the resulting solution was stirred at 400 rpm for the first 2 min, and then at 170 rpm. At pertinent intervals, an aliquot of the aqueous phase was sampled, and the concentration of Pb(II) in the sample was measured. From a decrease in the Pb(II) concentration in the aqueous phase, the uptake of Pb(II) at the sampled time was calculated.

2.5   Column-mode study

After an exchanger (0.4 g) was swollen with water, it was packed into a glass column (i.d. 7 mm) and the column was washed with water. The volume of the fibrous exchanger bed was 1.5 ml, which was used as the standard exchanger bed volume to convert flow rates in ml/h into space velocity in h^{-1}. The feeding solution was an aqueous solution of lead nitrate (0.01 M). After the feeding solution (113 - 157 bed volumes) was supplied to the column at a given space velocity, the column was washed with water.

Then, Pb(II) adsorbed was eluted by supplying 1 M nitric acid to the column at 3 h^{-1} in space velocity. After elution operation, the column was washed with water for next adsorption operation. All solutions and water were down flow supplied to the column by a peristaltic pump. All column effluents including washing were collected on a fraction collector. The concentration of Pb(II) in each fraction was measured in order to depict breakthrough and elution curves of Pb(II). In this work, breakthrough point of Pb(II) was designated as the volume of the feeding solution until C/C_0 attained to 0.05. Here, C and C_0 represent concentration of Pb(II) in the column effluent and the feeding solution.

3.   RESULTS AND DISCUSSION

3.1   Graft polymerization

In functionalization processes of precursory fibers prepared by electron pre-irradiation induced liquid phase graft polymerization technique, peeling of grafted polymers from trunk fibers and/or destruction of trunk PPPE fibers themselves becomes marked when the dg value of precursory grafted fibers exceeds 150.[11,12] On the other hand, precursory fibers with a dg value of 100 were successfully functionalized without such problems.[11,12] Then, the reaction time to give dg of 100 was searched. It took 2 h for PPPE-c and 5 h for PPPE-f in order to obtain CMS and ST grafted fibers with dg value of ca. 100 under the conditions described in the experimental section. Values of dg for CMS and ST grafted PPPE-c and PPPE-f used in the functionalization were 102 and 95, respectively, and respective chlorine contents were 2.8 and 2.4 mmol/g. It is interesting to clarify the cause of difference in the graft polymerization efficiency between the two trunk polymers PPPE-c and PPPE-f. However, no further work to clarify this phenomenon was performed, since the present work aimed development of bifunctional fibers and evaluation of metal ion adsorption abilities of the resulting fibrous cation exchangers.

3.2   Functionalization

Table 1 summarizes a few properties of the resulting fibers used in studies on their metal ion adsorption abilities. Hereafter, bifucntional fibers derived from PPPE-c and PPPE-f are denoted by symbols FPS-c and FPS-f, respectively, and respective symbols FP-c and FP-f denote monofunctional phosphonic acid fibers derived from PPPE-c and PPPE-f. Since phosphonic acid and sulfonic acid groups fixed on polymer matrices act as diprotic and monoportic acids, respectively, [6] it is expected that acid capacities of bifunctional fibers FPS are equal to sum of sulfur contents and twice phosphorus contents, and those of the monofucntional fiber FP and resin Diaion CRP200 are equal to twice phosphorus contents. Observed acid capacities approximately satisfy this expectation and support chemical structures of each exchangers depicted in Scheme 1. PPPE-f gave acid capacities higher than PPPE-c. This may indicates that the thinner fiber PPPE-f (0.9 denier) is more suitable for the functionalization than the thick fiber PPPE-c (1.5 denier).

Table 1. Properties of bifunctional and monofucntional fibers.

Exchangers	Phosphorus content (mmol/g)	Sulfur content (mmol/g)	Acid capacity (meq/g)
Bifunctional fiber			
FPS-c	0.91	1.28	2.91
FPS-f	1.58	1.30	4.56
Monofucntional fiber			
FP-c	1.20	-	2.66
FP-f	1.76	-	3.53
Cf. Diaion CRP 200	4.56	-	8.67

3.3   Batchwise study

Of interest is the metal ion selectivity of bifunctional exchangers having both sulfonic and phosphonic acid groups, since phosphonic acid resins prefer Pb(II) to Ba(II) but sulfonic acid resins show the reversed metal ion selectivity.

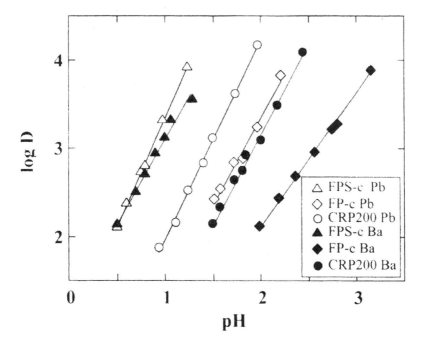

Figure 1. Log D vs. pH plots for Pb(II) and Ba(II). Ion exchanger 0.04 g; solution 25 ml of 1.0 x 10-4 M solution of Pb(II) or Ba(II); equilibrated for 24 h at 30 oC.

First, pH dependency of distribution ratios of Pb(II) and Ba(II) was studied to determine their half extraction pH values ($pH_{1/2}$), since $pH_{1/2}$ gives semi-quantitative measure of the metal ion selectivity. Figure 1 shows results for the three exchangers, FPS-c, FP-c, and Diaion CRP200, and Table 2 summarizes least square slopes of log D vs. pH plots and half extraction pH values. Slopes are nearly equal to 2 except for 1.4 of Ba(II)-FP-c system. At the present, however, the cause of this lower slope of is not clear. Values of $pH_{1/2}$ listed in Table 2 clearly suggest that FPS-c exhibits the highest affinity to both Pb(II) and Ba(II) among the tested three exchangers and slightly prefers Pb(II) to Ba(II).   On the other hand, monofunctional phosphonic acid exchangers FP-c and Diaion CRP200, which much more prefer Pb(II) to Ba(II) compared with FPS-c. Although both Diaion CRP200 and FP-c have phosphonic acid groups only, the former more strongly adsorbs the metal ions than the latter. This may be ascribable to the difference in densities of phosphonic acid groups between FP-c and Diaion CRP200.   The results in Fig. 1 and Table 2 suggest that the metal ion selectivity of FPS-c comes from cooperative recognition of Pb(II) and Ba(II) by both phosphonic and sulfonic acid groups.

Table 2. Least square slopes of log D vs. pH plots and values of half extraction pH (pH1/2)

	FPS-c		FP-c		Diaion CRP200	
	Pb(II)	Ba(II)	Pb(II)	Ba(II)	Pb(II)	Ba(II)
Slope	2.4	1.8	2.0	1.4	2.3	2.0
$pH_{1/2}$	0.78	0.84	1.71	2.38	1.37	1.67

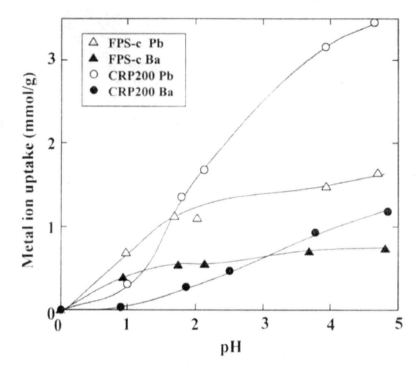

Figure 2. Capacities of ion exchangers for uptake of Pb(II) and Ba(II). Diaion CRP200 0.5 g, FPS-f 0.1g; solution 1.0 x 10-2 M solution of Pb(II) or Ba(II); equilibrated for 24 h at 30 oC.

Next, pH profiles of capacities were studied by using FPS-c and Diaion CRP200. Figure 2 gives the results. Above pH 2, Diaion CRP200 gave capacities for Pb(II) higher than those of FPS-c. Around pH 5, for instance, the capacity of Diaion CRP200 for Pb(II) is as high as 3.5 mmol/g, which is greater than twice that of FPS-c (1.6 mmol/g). However, the situation is reversed around pH 1, since sulfonic acid groups have much higher affinity to Pb(II) than phosphonic ones in low pH region. Interesting results can be seen in capacities for Ba(II). Marked difference is not observed in capacities for Ba(II) between Diaion CRP200 and FPC-c opposed to the case of Pb(II). This can be ascribed to the fact that sulfonic acid groups prefer Ba(II) to

Pb(II) but phosphonic acid groups does not prefer Pb(II). As shown here, the bifunctional FPS-c shows characteristic pH profiles in uptake of Pb(II) and Ba(II) different from those of monofunctional exchangers containing only either of phosphonic acid and sulfonic acid groups.

Last, kinetic performances of the three exchangers in Pb(II) uptake were evaluated. Figure 3 shows the results. FPS-c and FP-c gave the time independent uptake after 10 min and 20 min, respectively, whereas the granular resin Diaion CRP200 needed prolonged time of ca. 60 min to attain time independent uptake. Then, the increasing order of kinetic performances is Diaion CRP200 < FP-c < FPS-c.

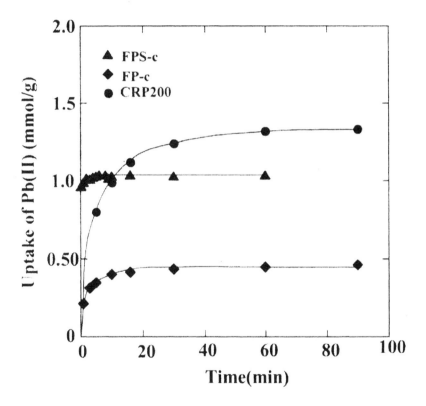

Figure 3. Time course for uptake of Pb(II) by FPS-c, FP-c, and CRP200. Resin or fibers 0.3 g; solution 150 ml of 5 x 10-3 M lead nitrate solution; temperature 30 oC.

3.4   Column-mode study

Since the batchwise study clarified that the bifunctional fiber FPS-c adsorb Pb(II) more rapidly than monofunctional fiber FP-c, of interest is the

dependence of breakthrough profiles of Pb(II)_on flow rates of the feed in
column-mode adsorption of Pb(II) by FPS-f and FT-f. These short cut
fibrous exchangers are suitable for homogeneous packing into columns
compared with the cloth type ones, which are difficult to pack into column
homogeneously. Then, column-mode adsorption of Pb(II)_was examined by
using FPS-f and FP-f packed columns. Indeed, smooth feeding of solutions
to FPS-f and FP-f packed columns were achieved up to a high space velocity
of 900 h-1, which is 90 times faster flow rates adopted in granular resin
packed column.

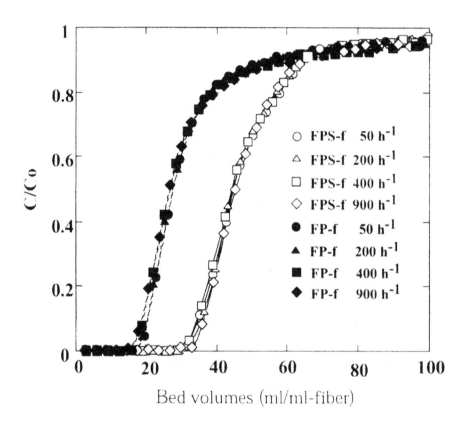

Figure 4. Breakthrough curves of Pb(II) in the adsorption of Pb(II) by FPS-f and FP-f packed
columns. Column 1.5 ml of wet fiber (0.4 g in dry state), feeding solution 0.01 M lead nitrate.
Flow rates in space velocity (h-1) are denoted on the figure.

Figure 4 shows breakthrough profiles during the adsorption operation of
Pb(II) by both FPS-f and FP-f packed columns. Surprisingly, both columns
gave flow rate independent breakthrough profiles of Pb(II), although the

slight difference in adsorption kinetics was observed in the case of batchwise approach. Thus, much faster flow rates will be needed to observe flow rate dependent breakthrough profiles. However, the flow rate of 900 h^{-1} is nearly equal to the upper limit of our present column system.

Lead(II) adsorbed on both columns was quantitatively eluted with ca. 6 bed volumes of 1 M nitric acid. In this work, 2 cylces of the adsorption-elution operation were repeated at each flow rate. Then 16 cycles of the adsorption and elution operation were conducted for each column. In the case of FPS-f column, averaged recovery was 104% with standard deviation (sd) of 4 % (n = 16) and for the other column, averaged recovery 103% with sd of 4 %. Table 3 summarizes numerical data for results shown in Fig. 4. Breakthrough capacities at C/Co = 0.05 are 0.54 – 0.57 and 0.28 – 0.33 mmol/g for FPS-f and FP columns, respectively. Total uptake was 0.78 – 0.83 mmol/g for FPS-f and 0.51 – 0.54 mmol/g for RF-f. During repeated adsorption-elution operations, no deterioration of both RFP-f and FP-f was observed.

Table 3. Performances of FPS-f and FP-f packed columns.

Exchanger	Flow rate of feed	Flow rate of feed	Flow rate of feed	5% Breakthrough capacity	Total uptake
	h^{-1}	(bed volumes)	(ml/ml-fiber)	(mmol/g)	(mmol/g)
FPS-f	50	149(10)[a]	33	0.55	0.83
	200	139(9.8)	34	0.56	0.81
	400	134(9.8)	33	0.54	0.78
	900	129(9.8)	35	0.57	0.79
FP-f	50	136(9.8)	20	0.33	0.54
	200	134(9.9)	19	0.31	0.54
	400	130(9.8)	17	0.28	0.52
	900	129(9.9)	18	0.30	0.51

[a] Figures in parentheses denote the concentration of Pb(II) in the feed (mM).

As conclusion, bifunctional fibers having both phosphonic acid and sulfonic acid groups exhibit the characteristic metal ion selectivity and high breakthrough capacities in addition to the extremely fast adsorption rates. Studies on behavior of FPS-f in adsorption of other heavy metals like Fe(III) are now in progress. The bifunctional fiber developed in this work is attractive to application to the protection of the environment because of its extremely rapid adsorption rates and characteristic metal ion selectivity.

Authors thank Seimi Chemical Co. Ltd. for a gift of chloromethylstyrene and Mitsubishi Chemical Co. Ltd. for a gift of Diaion CRP200. We also acknowledge Ms. Ayumi Meakawa for her help in measurement, and Dr. Shoji Aoki, Mr. Mitsuhiro Nakao for their help in graft polymerization.

REFERENCES

1. Jyo, A., Yamabe, K., Egawa, H., Behavior of macroreticular chelating resins having oxy acids of phosphorus in adsorption and elution of lead ion. in: Chemistry for the Protection of Environment, 2, eds. L. Pawlowsky, W. J. Lacy, C. G. Uchrin, M. R. Dudzinska, Environmental Science Research, Vol. 55, Plenum Press, New York, 1996, pp. 121-129, ISBN 0-306-45373-8.
2. Jyo A., Yamabe K., Egawa H., Behavior of methylenephosphonic acid chelating resin in adsorption and elution of molybdenum (VI), Sep. Sci. Technol. 31, 1996, 513-522.
3. Jyo A., Yamabe K., Egawa, H., Metal ion selectivity of macroreticular styrene divinylbenzene copolymer-based methylenephosphonic acid resin. Sep. Sci. Technol. 32, 1997, 1099-1105.
4. Yamabe, K., Ihara, T., Jyo, A., Metal ion selectivity of macroreticular chelating cation exchange resins with phosphonic acid groups attached to phenyl groups of a styrene-divinylbenzene copolymer matrix, Sep. Sci. Technol. 36, 2001, 3511-3528.
5. Egawa, H., Nonaka, T., Ikari, M., preparation of macroreticular chelating resins containing dihydroxyphosphino and/or phosphono groups and their adsorption ability for uranium. J. Appl. Polym. Sci. 29, 1984, 2045-2055.
6. Trochimczuk, A. W., Alexandratos, S. D., Synthesis of bifunctional ion-exchange resins through the arbusov reaction: effect on selectivity and kinetics, J. Appl. Polymn. Sci. 52, 1994, 1273-1277.
7. Alexandratos, S. D., Trochimiczuk, A. W., Horwitz, E. P., Catrone, R. C., synthesis and characterization of a bifunctional ion exchange resin with polystyrene-immobilized diphosphonic acid ligands, J. Appl. Polymn. Sci. 61, 1996, 273-278.
8. Chiariza, R., Horwitz, E. P., Alexandratos, S. D., Gula, M. J., ; diphonix resin: a review of its properties and applications, Sep. Sci. Technol. 36, 2001, 3511-3528.
9. Beauvais, R. A., Alexandratos, S. D., polymer-supported reagents for the selective complexation of metal ions: An overview. React. Funct. Polym. 36, 1998, 113-123.
10. Alexandratos, S. D., Natesan, S., Ion-selective polymer-supported reagent: principle of bifunctionality, Eur. Polym. J. 35,1999, 431-436.
11. Jyo, A., Aoki, S., Kishita, T., Yamabe, K., Tamada, M., Sugo, T., Phosphonic acid fiber for selective and extremely rapid elimination of lead(II), Anal, Sci. 17, 2001, i201-i204.
12. Aoki, S., Saito, K., Jyo, A., Katakai, A., Sugo, T., Phosphoric acid fiber for extremely rapid elimination of heavy metal ions from water, Anal, Sci. 17, 2001, i205-i208.
13. Jyo, A., Saito, K., Aoki, S.; unpublished work.
14. Maeda, H., Egawa, H., Preparation of macroreticular chelating resins containing mercapto groups from 2,3-epithiopropyl methacrylate/divinylbenzene copolymer beads and their adsorption capacity, Anal. Chim. Acta 162,1984, 339-346.
15. Zhu, X., Jyo, A., Column-mode removal of lead ion with macroreticular glycidyl methacrylate-divinylbenzene copolymer-based phosphoric acid resins, J. Ion Exchange 11, 2000, 68-77.

2

FATE OF CONTAMINANTS IN THE ENVIRONMENT

ARSENIC SPECIATION FOR INVESTIGATION OF ITS ENVIRONMENTAL FATE – A CASE STUDY

Birgit Daus[1*], Jürgen Mattusch[2], Rainer Wennrich[2], Holger Weiss[1]

[1]*UFZ-Umweltforschungszentrum Leipzig-Halle GmbH, Interdisziplinary Department Industrial and Mining Landscapes, Permoserstrasse 15, 04318 Leipzig, Germany, E-mail: daus@pro.ufz.de, weiss@pro.ufz.de; [2]UFZ-Umweltforschungszentrum Leipzig-Halle GmbH, Department of Analytical Chemistry, Permoserstrasse 15, 04318 Leipzig, Germany, E-mail: wennrich@ana.ufz.de, mattusch@ana.ufz.de*

Abstract: A tailings pond for dumping the fine grained residues of the tin ore processing was object of this study. The aim was the investigation of arsenic mobilization from the tailings material into the seepage water and the immobilization process into a precipitate. Different speciation techniques were applied to follow the way of the arsenic in this system. The seepage water was sampled and the total concentrations of the main constitutions as well as the arsenic species were analyzed. The mobilization and immobilization process was described by these methods.

Key words: Arsenic, environmental fate, speciation

1.   INTRODUCTION

Since medieval times tin ore has been mined the Erzgebirge region, in the southern part of Eastern Germany. A settling pond for the tailings was constructed isolating a segment of the Biela River valley near the city of Altenberg by two cross-valley dams, an upstream dam with a height of 26 m and a downstream dam with a height of 89 m. This was used for dumping the residues of tin ore processing. The pond covers an area of about 0.53 million m^2 and has a volume of about 10.6 million m^3 [1]. The tailings were deposited as a fine-grained (60-80 % <63 μm) slurry with a water content of about 85 %. To enable the consolidation of the tailings sediment, the surface water was drained and the downstream dam was constructed as a "seepage dam" with a seepage rate of 30-40 liters/second out of three different sources. This tailings pond was used between 1967 and 1991 for the deposition of slurries.

The aim of this work was to investigate the arsenic mobilization from the tailings material (200 – 500 µg/g As) into the seepage water (up to 3.5 mg/L As) and the process of seepage water effluent forming an immobilized precipitate (up to 8 % As) in the creek. Different analytical methods for the determination of total concentrations and different sequential extraction methods as well as hyphenated techniques for speciation analysis were applied to follow the way of the arsenic in this environment.

2.   EXPERIMENTAL

• Sampling

Solid samples of tailings material were collected from different depths (2 - 18 m) in drill cores. Precipitates were collected from different locations in the stream bed. Water samples were taken from a well situated in the tailings pond, at three different effluent points on the downstream face of the tailings dam, and also at 30 m as well as 50 m downstream. The physico-chemical parameters of the water were measured during sampling.

The samples were collected in polyethylene bottles and stored at 4°C. Water samples were stabilized for arsenic speciation using phosphoric acid to yield a final concentration of 10 mM [2]. Wet material was used for the extraction and the dry weight was calculated by weighing a different part of the sample after drying (105°C).

• Analysis

The determination of the total concentrations of metal ions and arsenic in the water samples and in the eluates of solid materials were carried out using ICP-AES (Spectroflame, SPECTRO A.I.) with pneumatic nebulization (cross flow). Anion (SO_4^{2-}, Cl^-) determinations were done using an ion chromatographic device with IonPac AS12A/AG12A column and a conductivity detector.

A hyphenated technique, the coupling of ion chromatographic separation (DX-100, DIONEX) and inductively coupled plasma mass spectrometric detection (ELAN-5000, PERKIN-ELMER) was used for the speciation analysis of As(III) and As(V) [3]. After 2001 another ICP-MS instrument (PQ ExCell; THERMO) was used for the coupled method.

- ## Sequential extraction of tailings material

The sequential extraction scheme developed by Foerstner [4] was chosen for these investigations except for one modification. In the last step a digestion with *aqua regia* (German Regulations DIN 38414, part 7) was used.

- ## Sequential extraction of the precipitate samples

This special method was used for iron hydroxides precipitates to distinguish arsenic adsorbed to the hydroxides, co-precipitated with them or bound in minerals is described in Daus *et al*. [5].

3.   RESULTS

The investigation of the tailings material has shown a very homogeneous element pattern over the depth. The total arsenic concentration is 270 ± 40 mg/kg. The main binding form of the arsenic in the tailings material determined by the sequential extraction method is partly as sulfides and partly bound to the amorphous or poorly crystallized iron hydroxide phases. The latter is about 44 ± 8 % of total arsenic found in samples from different depths from a drilling core. Taking into the account the measured redox and pH values in some wells inside the tailings pond as well as the conditions at the seepage water source, a reduction of the amorphous or poorly crystallized iron hydroxides seems to be the main mobilization process for arsenic.

There are three sources of seepage water out of the dam. These ones and additional sampling points were 30 m and 50m downstream were selected for sampling. The characteristic measures of the seepage water are summarized in Table 1. Only the middle source seems to be in contact with the tailings material due to the high arsenic and iron values and the low redox potential. The neutral pH and the relatively low sulphate concentrations are in agreement with the hypothesis of an reductive mobilization and disagree with the typical values for acid mine drainage (e.g. [6]). Also tri-valent arsenic is present in a great proportion, indicative of a reductive mechanism. The speciation of arsenic shows different proportions

in dependence on a filtration (0.45 μm, cellulose acetate) of the water during
sampling (c.f. Figure 1). The forming of fine-grained iron hydroxide seems
to have a high influence on the binding species and the concentration of
arsenic in the water, as expected. The arsenate is bound to small and mobile
particles to a high extent.

Samples taken 30 m or 50 m downstream show still high concentrations
of arsenic as well as iron (Table1). Additionally, the arsenite concentration is
remarkably high (about 30 % of the total As in an unfiltered sample).

Table 1. Measured pH values, Eh [mV], oxygen concentrations [mg/L] and water constituents
[mg/L] at the different sampling sites.

	left spring	middle spring	right spring	30m downstream	50m downstream
pH	6.1	6.3	6.0	6.4	6.6
Eh	211	202	202	183	190
O2	6.6	2.1	11	6.3	6.5
As	0.05	3.3	0.06	2.5	2.5
Fe	0.03	10.9	0.03	7.9	8.1
Mn	0.3	7.9	0.08	5.1	5.9
SO42-	43.8	82.1	87.8	85.8	85.3
Cl-	3.47	11.3	2.98	8.4	8.4

Experiments were performed in the laboratory to investigate the
oxidation process under the given conditions [7]. The oxidation of
arsenic(III) was found to be slow and incomplete. After about 100 hours a
steady state on a level of about 40 % of the starting concentration was
determined.

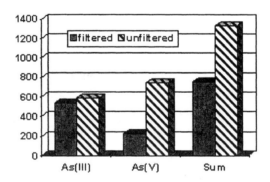

Figure 1. Differences between filtered and unfiltered samples from the seepage water

A precipitation of the iron hydroxides is also observed. The reddish precipitate in the creek starts right away at the source and contains up to 8 %(w/w) of arsenic. As shown in a previous publication [5] the arsenic is bound only weak to the iron hydroxides. A mobilization is possible by desorption without dissolution of the iron hydroxides.

4. CONCLUSIONS

The analytical methods described were found to be suitable for the investigation of the fate of arsenic in this environmental system.

The mobilization of arsenic from the tailings material seems to be a slow and continuos process attributed to reduction of iron phases. The seepage water of the middle source contains arsenite as well as arsenate in high concentrations and seems to be the only water source in contact with the tailings material. The concentrations of arsenic downstream are still high and the immobilization process by precipitation of iron hydroxide and co-precipitation or sorption of arsenic is incomplete. A reason for this may be the slow kinetics of the oxidation process and the transport of fine grained hydroxide particles. These particles are mobile and can bind the arsenic (mainly as arsenate) too.

The immobilized arsenic in the precipitate is bound only by sorption onto the amorphous iron hydroxides. A sustainable immobilization would need additional action.

REFERENCES

1. H. Weiss, B. Daus, J. Mattusch (1999): Arsenic mobilization and precipitation from a tin mill tailings in the Erzgebirge, Germany.- Chron. Rech. Min. 534, 13 - 19
2. B. Daus, J. Mattusch, R. Wennrich, H. Weiss (2002): Investigation on stability and preservation of arsenic species in iron rich water samples.- Talanta, in press
3. J. Mattusch and R. Wennrich (1998): Determination of anionic, neutral, and cationic species of arsenic by ion chromatography with ICPMS Detection in environmental samples.- Anal. Chem. 70, 3649 - 3655
4. U. Förstner (1993): Metal Speciation – General Concepts and Applications.- Intern. J. Environ. Anal. Chem. 51, 5 - 23
5. B. Daus, H. Weiss, R. Wennrich (1998): Arsenic speciation in iron hydroxide precipitates.- Talanta 46, 867 – 873
6. L. Carlson, J.M. Bigham, U. Schwertmann, A. Kyek, F. Wagner (2002): Scavenging of As from Acid mine Drainage by Schwertmannite and Ferrihydrite: A comparison with synthetic analogues.- Environm. Sci. Technol., preprint

CHEMICAL SPECIATION OF ALUMINUM IN DELHI SOILS

Swarna Muthukrishnan and D. K. Banerjee
School of Environmental Sciences, Jawaharlal Nehru University, New Delhi, 110 067, India;
16 Rachel Court, Kendall Park, NJ 08825 USA, E-mails: swarna_m@yahoo.com;
dkb0400@mail.jnu.ac.in

Abstract: The objective of the investigation was to ascertain the status of Aluminum in Delhi soils by studying its speciation in the soil profile and to assess if there was any spatial variability. Soils representing the Aravali Ridge and the alluvial floodplains of river Yamuna were collected as a single, undisturbed core up to a depth of 1m and the profile differentiated into four layers- 0-17 cm, 17-37 cm, 37-57 cm, and 57-86 cm. "Pseudo" total Aluminum and Iron in the soils were speciated into the "operationally" defined species (weakly exchangeable, organic matter complexes, amorphous oxides and hydroxides, and crystalline or "free" oxides) by widely recommended selective extraction procedures. Both Al and Fe in these soils are bound predominantly as Fe oxides and silicates and have only very low percentages in the easily mobilizable pools.

Key words: Solid-phase Al, soils, mobility, chemical speciation, selective extractions, Al_{pstot}, Al_{Cu}, Al_{py}, Al_{ox}, Al_{DCB}, Fe_{pstot}, Fe_{py}, Fe_{ox}, Fe_{DCB}

1. INTRODUCTION

Aluminum occurs in many forms in soils, and its chemistry is most complicated [1, 2, 3, 4, 5]. Various solid-phase Al pools occur in soils, differing in molecular structure, intralattice bond energies, hardness and surface charge. There are significant differences in amounts and forms of solid-phase Al in soils, depending mostly on age of soil, parent material, climate and topography. Theoretically, Al in soils can be physically divided into:

(i)   crystalline Al (primary and secondary minerals);

(ii)   non-crystalline Al (such as amorphous Al oxides and hydroxides);

(iii)   organic-bound Al (Al-humus complexes);

(iv)   interlay Al and polymer Al;

exchangeable Al (specifically and non-specifically adsorbed Al)[6].

Long-term effects on aqueous aluminum in soils, resulting from changes in the soil solid phase, are generally ignored. The rate of leaching of an appreciable portion of the various solid-phase aluminum pools under acidic soil conditions could lead to a significant depletion of certain soil aluminum fractions within several decades. The current depletion rates of extractable aluminum are significantly higher than the rates of aluminum transfer from silicate-bound to extractable soil aluminum which may have serious implications on soil-forming processes, the health of plant communities and clay mineralogy.

An earlier study on the speciation of heavy metals and geochemical mapping of the total metal content of the surface soils of Delhi[7] showed the total Al content to vary between 1.87 to 5.34% with a mean of 3.57%. In view of the above, the objective of this research was to ascertain primarily the status of Aluminum in Delhi soils by studying its chemical speciation in the soil profile in order to improve the understanding of its distribution in the solid-phase pool of the soil. More precisely, the objectives were:

1. To study the chemical speciation of Aluminum in the solid – phase of the selected soil samples by selective chemical extraction procedures,
2. To examine the vertical distribution of the different Aluminum species in the soil profile and determine any spatial variability in the Al species between the ridge and the floodplain soils,
3. To assess the influence of some soil physico-chemical characteristics, viz., pH, organic matter, clay content, etc., and exchangeable cations, on the spatial variation and vertical distribution of the solid-phase Al species in the profile of Delhi soils at selected sites.

2.   STUDY AREA

The Delhi region (28°23'17" and 28°53'00"N, 76°50'24" and 77°20'37"E) is a part of the Indo-Gangetic alluvial plains, situated at an elevation ranging from 198 to 220 m above msl, and occupies an area of 1483 Km^2. The land of Delhi comprises the broken ridge of the Aravali Mountains and the flood plains of the river Yamuna. The Delhi Ridge, which is the culminating spur of the Aravalis constitutes the most significant physiographic feature of the region. It enters Delhi from the south and extends into a northeasterly direction. The ridge forms the principal watershed of the area and acts as ground water divide between the western and eastern parts of the Delhi area. The only perennial river, Yamuna flows in southerly direction in the eastern extremity of the Union Territory. The soils of Delhi are alluvial in nature and have been influenced by the flow of river Yamuna and its floodwaters, the ridge and the wind blowing from the southwest. These soils are generally low in available nitrogen, low to medium in phosphorus, medium to high in potassium, adequate in calcium, magnesium and sulphur. The soils can be classified into four broad divisions – *Khadar (*new alluvium), *Bhangar* (old alluvium), *Dabar* (found in low-lying areas), and *Kohi* (rocky areas)[8].

3.   MATERIALS AND METHODS

For the Aluminum speciation study, soils were selected from 11 sites comprising the Delhi Ridge and the floodplains of the river Yamuna as shown in Figure 1. Five of these (P6, L14, L20, J24, J26) represented the broken Ridge of the Aravalis in the north, central and southern part of Delhi, and six samples (D8, L4, R4, X10, X14, and X20), were from the Yamuna floodplains of Delhi. Each soil sample was collected as a single, uniform, undisturbed core up to a depth of 1m on a 2km x 2 km grid, using a specially designed stainless steel sampler in order to study the speciation in the whole profile. The speciation was studied by classifying the soil profile into four different layers: 0-17cm, 17-37cm, 37-57cm and 57-86cm, respectively. Representative soil samples of the different soil layers (<2 mm) were used for the investigation. All chemicals used in this study were of analytical grade, and de-ionized water was used during the analyses. All extractions were conducted in triplicate in acid-washed (10% HNO_3) polycarbonate and borosilicate labware. Multi level standards for all metals were prepared for each extraction step in the same matrix as the extracting reagents to minimize matrix effects. Blanks were used for background correction and other sources of error. The precision of the method was verified by running one duplicate for every 10 samples analyzed.

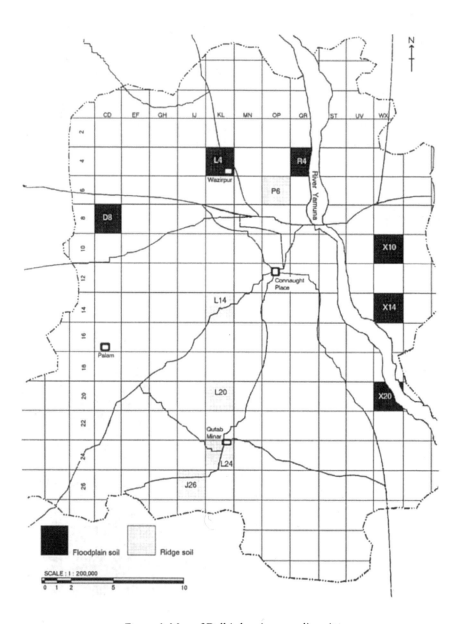

Figure 1. Map of Delhi showing sampling sites

Physico-chemical parameters investigated included pH, electrical conductivity (1:2.5, soil:H_2O), TOM (Walkley-Black Method), and silt and clay content (Stoke's method). Exchangeable cations, K, Na, Mg, Ca, Fe and Mn, were determined by extraction with 0.1M $BaCl_2$ and subsequent measurement by Atomic Absorption Spectrometry. The soils in the 0-17cm, 17-37cm, 37-57cm, and 57-86cm layers of the profile, were subjected to separate extraction procedures involving specific reagents, to give the following six "operationally defined" species of Al:

(i)   "pseudo" total Al, Al_{pstot}

(ii)   (Exchangeable Al, Al_{ex} [9]

(iii)   Weakly-bound organic complexes of Al, Al_{Cu} [10]

(iv)   Strongly-bound organic complexes of Al, Al_{py} [11]

(v)   Amorphous oxides and hydroxides of Al, Al_{ox} [12]

(vi)   Crystalline or "free" oxides and hydroxides of Al, Al_{DCB} [13].

Iron in the soil samples was also speciated to yield the following "operationally defined" species of Fe:

(i)   pseudo" total (acid-reactive) Iron, Fepstot

(ii)   Strongly-bound organic complexes of Fe, Fepy

(iii)   Amorphous oxides and hydroxides of Fe, Feox

(iv)   Crystalline or "free" oxides and hydroxides of Fe, FeDCB.

4.   RESULTS AND DISCUSSION

Descriptive statistics of the soil physico-chemical characteristics and exchangeable cations are shown in Table 1. The mean concentrations of the "operationally" defined species of Al and Fe in the ridge and floodplain soils are shown in Table 2.

0-17cm	pH	EC	(silt + clay)	clay	OM	ex-K	ex-Na	ex-Mg	ex-Ca	ex-Fe	ex-Mn	Σ cations
		(mS/cm)	(%)						(mg/100 g)			
mean	8.13	0.61	28.79	16.92	2.01	27.45	11.39	25.78	82.12	1.44	0.66	140.58
SD	0.20	0.20	6.34	5.20	0.58	5.29	6.77	2.83	11.30	0.45	0.17	17.84
meanridge	8.14	0.53	28.10	15.91	2.14	30.12	9.35	26.02	86.27	1.53	0.60	153.89
SD	0.26	0.13	6.26	5.53	0.74	6.28	7.44	3.66	7.56	0.27	0.16	15.92
meanfp	8.13	0.68	29.36	17.77	1.89	25.23	13.09	25.57	63.33	1.56	0.70	129.48
SD	0.16	0.23	6.95	5.26	0.46	3.38	6.30	2.29	10.27	0.15	0.17	10.45

17-37cm	pH	EC	(silt + clay)	clay	OM	ex-K	ex-Na	ex-Mg	ex-Ca	ex-Fe	ex-Mn	Σ cations
		(mS/cm)	(%)						(mg/100 g)			
mean	8.30	0.51	23.10	14.28	0.79	14.20	18.07	25.44	75.77	2.10	0.22	127.48
SD	0.16	0.17	6.58	5.04	0.29	2.61	8.54	3.57	13.03	0.65	0.09	18.12
meanridge	8.23	0.47	23.11	13.97	0.82	13.98	17.48	24.26	80.48	2.26	0.21	138.67
SD	0.20	0.18	7.24	6.18	0.40	3.10	12.99	4.17	12.41	0.71	0.08	21.22
meanfp	8.36	0.54	23.09	14.53	0.76	14.39	18.56	26.12	56.61	1.96	0.22	118.16
SD	0.10	0.17	6.68	4.47	0.19	2.41	3.16	3.00	6.98	0.63	0.10	8.19

37-57cm	pH	EC	(silt + clay)	clay	OM	ex-K	ex-Na	ex-Mg	ex-Ca	ex-Fe	ex-Mn	Σ cations
		(mS/cm)	(%)						(mg/100 g)			
mean	8.30	0.59	33.91	14.88	0.68	12.00	13.69	26.07	89.31	1.01	0.36	130.90
SD	0.15	0.21	9.89	5.07	0.23	3.11	9.43	7.10	20.38	0.15	0.07	22.10
meanridge	8.20	0.52	31.41	12.94	0.68	10.99	11.42	21.21	96.51	1.09	0.35	141.58
SD	0.09	0.20	6.23	5.34	0.32	4.15	10.97	4.27	19.43	0.18	0.10	22.88
meanfp	8.38	0.64	36.00	16.50	0.68	12.84	15.58	30.12	62.15	0.95	0.37	122.01
SD	0.13	0.22	12.38	4.66	0.14	1.93	8.49	6.56	9.07	0.10	0.04	18.69

57-86cm	pH	EC	(silt + clay)	clay	OM	ex-K	ex-Na	ex-Mg	ex-Ca	ex-Fe	ex-Mn	Σ cations
		(mS/cm)	(%)						(mg/100 g)			
mean	8.48	0.63	33.52	16.16	0.56	13.72	21.36	26.54	85.33	1.84	0.23	137.95
SD	0.26	0.21	8.87	7.15	0.31	4.40	15.36	9.54	12.92	0.17	0.12	25.92
meanridge	8.31	0.57	35.40	16.95	0.49	13.06	11.36	24.33	88.53	1.79	0.18	139.25
SD	0.14	0.31	11.97	9.03	0.24	5.33	9.16	11.30	12.05	0.20	0.07	20.89
meanfp	8.62	0.69	31.96	15.50	0.62	14.27	29.69	28.37	62.38	1.88	0.27	136.87
SD	0.26	0.10	6.02	5.98	0.38	3.90	14.87	8.42	13.67	0.15	0.15	31.49

mean – mean values of all soil samples, mean ridge – mean values of ridge soils, mean fp – mean values of flood plain soils

Table 1. Physico-chemical characteristics and exchangeable cations in the soil profile

0-17cm	Al_{pstot}	Al_{Cu}	Al_{py}	Al_{ox}	Al_{DCB}	Fe_{pstot}	Fe_{py}	Fe_{ox}	Fe_{DCB}
mean	11261.6	98.11	142.78	202.78	157.65	14692.1	176.11	532.80	1608.93
SD	1473.4	34.62	37.44	69.73	8.85	3319.2	31.79	111.88	103.20
meanridge	11723.0	99.30	165.06	191.58	159.27	12529.8	181.93	485.21	1672.66
SD	1019.6	30.84	41.98	93.08	2.41	1462.5	22.39	133.92	25.08
meanfp	10877.0	97.12	124.22	212.11	156.30	16494.0	171.27	572.45	1555.82
SD	1766.3	40.42	22.01	50.63	12.13	3428.1	39.47	80.84	115.54
17-37cm	Al_{pstot}	Al_{Cu}	Al_{py}	Al_{ox}	Al_{DCB}	Fe_{pstot}	Fe_{py}	Fe_{ox}	Fe_{DCB}
mean	12721.2	82.05	134.21	151.26	148.73	15201.1	170.15	432.98	1622.64
SD	1238.1	8.81	19.93	67.01	10.76	4775.9	67.63	190.76	175.89
meanridge	12201.9	84.90	136.02	141.90	147.45	11569.5	152.30	407.88	1698.09
SD	974.6	7.13	19.81	51.51	11.40	1832.4	38.24	138.62	85.48
meanfp	13154.0	79.67	132.69	159.07	149.79	18227.5	185.02	453.90	1559.77
SD	1346.0	9.99	21.79	81.84	11.17	4330.5	85.99	237.17	213.53
37-57cm	Al_{pstot}	Al_{Cu}	Al_{py}	Al_{ox}	Al_{DCB}	Fe_{pstot}	Fe_{py}	Fe_{ox}	Fe_{DCB}
mean	14001.9	78.13	115.23	40.69	140.39	15996.4	139.34	183.53	1541.01
SD	2117.0	6.92	25.54	23.57	22.74	5349.0	33.88	57.18	306.00
meanridge	14333.6	79.12	115.30	28.40	136.08	12079.4	154.70	160.27	1433.72
SD	2628.3	8.31	35.20	10.10	12.76	2314.6	35.43	45.36	217.38
meanfp	13725.6	77.30	115.17	50.92	143.99	19260.5	126.55	202.92	1630.43
SD	1798.6	6.22	17.72	27.44	29.49	4981.0	29.31	62.46	358.28
57-86cm	Al_{pstot}	Al_{Cu}	Al_{py}	Al_{ox}	Al_{DCB}	Fe_{pstot}	Fe_{py}	Fe_{ox}	Fe_{DCB}
mean	13309.6	86.10	109.14	213.95	141.03	16216.5	178.12	445.29	1440.51
SD	1685.3	15.37	26.46	137.21	8.12	5732.7	86.75	228.67	206.34
meanridge	12925.0	87.77	117.58	232.21	144.26	11375.9	157.02	471.03	1476.84
SD	2286.8	13.39	26.45	155.81	6.25	1270.8	40.24	224.22	117.46
meanfp	13630.0	84.70	102.10	198.74	138.35	20250.4	195.70	423.85	1410.23
SD	1107.1	17.99	26.64	132.74	9.03	4634.5	113.75	251.29	267.76

mean – mean values of all soil samples; mean ridge – mean values of ridge soils, mean fp – mean values of flood plain soils

Table 2. Descriptive statistics of Al and Fe species in the soil profile (all concentrations in ppm)

5. AL AND FE SPECIES IN THE SOIL PROFILE

"Pseudo" total Aluminum concentration was the lowest in surface soils (1.13 ± 0.15%) and reached a maximum in 37-57cm depth (1.4 ± 0.21%). The increase in the concentration of Al_{pstot} with depth upto 57cm implied a maximum accumulation in 37-57cm depth. Al_{pstot} in surface layers appeared to be independent of the influences of soil physical characteristics, exchangeable cations, and the different species of Al and Fe. The contribution to Al_{pstot} from the weakly exchangeable(Al_{Cu}), organic-bound complexes(Al_{py}) and amorphous oxides and oxyhydroxides(Al_{ox}) reached a minimum in 37-57 cm (Figure. 2a-d), which suggests that much of Al_{pstot} may be bound strongly to silicates, and hence not available for speciation under the specified conditions. The decrease in Al_{py} with depth with a corresponding decrease in soil organic matter implied that the speciation of Al as Al_{py} depends very much on the availability of organic matter for complexation. The contribution of Al_{DCB} to Al_{pstot} content was more or less uniform in all the four layers of the soil profile of all the soils investigated. While amorphous oxides dominated the speciation in the surface layer, Al_{DCB} was the most abundant species in depths upto 57 cm. While the maximum contribution was by amorphous and crystalline oxides in the various depths, weakly-exchangeable complexes accounted the least towards speciation on account of the alkaline soil pH conditions. Fe_{pstot} in these soils varied between 0.86 and 2.46%, and showed a greater abundance with depth of the soil profile. Thus, the 57-86cm layer exhibited the highest Fe_{pstot} (1.62 ± 0.57%), and the 0-17cm layer, the lowest Fe_{pstot} (1.47 ± 0.33%) concentrations respectively. The entire soil profile was dominated by crystalline Fe oxides, followed by amorphous oxides, while the organic bound complexes of Fe were the least abundant species.

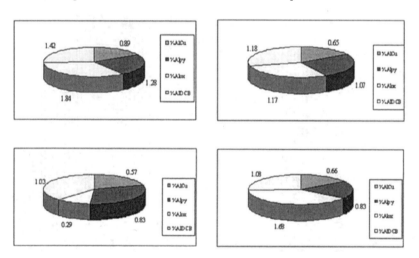

Figure 2. a-d. Distribution of Al species in 0-17cm, 17-37cm, 37-57cm, 57-86cm in the soil profile

6. VERTICAL DISTRIBUTION OF AL SPECIES

The concentrations of the four Al species occurring in each layer of the soil profile are expressed as a percentage of the Al_{pstot} content of the specific soil layer. While Al_{pstot} increased with depth of the profile, the contribution of the operationally defined Al species towards speciation however decreased with depth (Figure 3a-d). The maximum differentiation of Al_{pstot} as Al_{Cu}, Al_{py}, Al_{ox}, and Al_{DCB} was 8.51% which was observed in the surface layer while the mean contribution was 5.43 ± 1.60%. On the contrary, in 37-57 cm depth which had the most abundant Al_{pstot}, the mean share of the Al species (2.72 ± 0.75%) was the lowest contribution to Al_{pstot} in all the layers of the soil profile. This implied that "pseudo" total Al in these depths are predominantly bound as silicates and hence are not available for speciation under the experimental conditions.

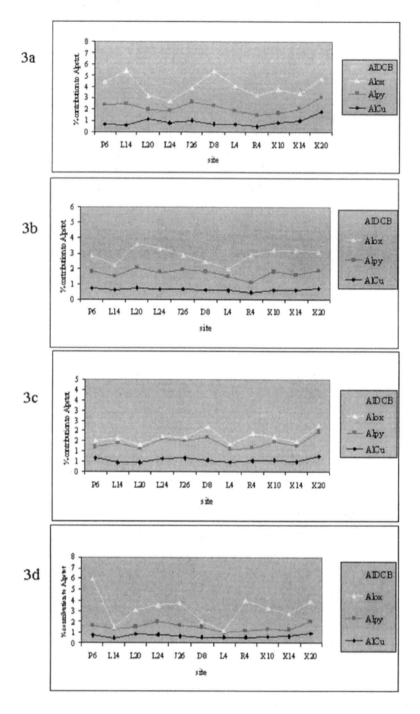

Figure 3.a-d "Operationally" defined Al species in 0-17cm, 17-37cm, 37-57cm, 57-86cm in the soil profile

7. CORRELATION STUDIES

Correlations in the Al species were assessed and their significance tested using two-tailed tests [14] for all soil samples in all the four layers, i.e. 0-17, 17-37, 37-57, and 57-86cm, respectively. The 0-17cm was characterized by a positive correlation between pyrophosphate-extractable Al and the total organic matter of the soil ($r = 0.733$, $p \leq 0.05$). The strong and significant relations between ex-K ($r = 0.788$, $p \leq 0.01$), ex-Ca ($r = 0.802$, $p \leq 0.01$) and Al_{py} (Figure 4) implied the existence of these cations and Al bound to soil organic matter and extraction with an unbuffered salt resulted in the dissociation of these metal ion-organic matter complexes. Soil pH greater than 8.0 in the surface layer decreased the occurrence of amorphous oxides and hydroxides of Al ($r = -0.936$, $p \leq 0.01$) (Figure 5) and apparently increased the solubility and exchangeability of the weakly held Al complexes, resulting in an increase in Al_{Cu} species. Al_{Cu} was negatively correlated to ex-Mn in 0-17cm in the soil profile ($r = -0.749$, $p \leq 0.01$) (Figure 6) which was in direct contrast to that of Al_{ox} ($r = 0.770$, $p \leq 0.01$) (Figure 7). However, the positive "r" values between ex-Fe and Al_{Cu} ($r = 0.720$, $p \leq 0.05$), and between ex-Fe and Al_{py} ($r = 0.695$, $p \leq 0.05$), in 17-37 cm could be due to the possible dissolution of ex-Fe and Al_{Cu} from their organic bound complexes. This was further reinforced when it was observed that both Al_{Cu} and Al_{py} were positively correlated ($r = 0.770$, $p \leq 0.05$). However, with further increase in depth, the soil physical characteristics did not play a significant role in influencing the speciation of Al_{pstot} and Fe_{pstot}. In addition, the number of species of Al and Fe, being influenced by existing physical conditions and the chemical nature of soils also decreased on moving down the soil profile. Positive correlations between amorphous oxides and hydroxides of Al and Fe ($p \leq 0.01$) and crystalline and free oxides of Al and Fe ($p \leq 0.05$) indicated a common crustal occurrence.for these species in lower depths in these soils

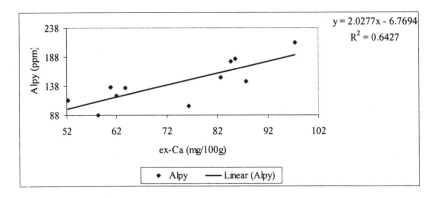

Figure 4. Alpy vs. ex-Ca in 0-17cm layer of the soil profile

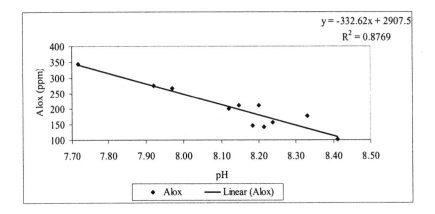

Figure 5. Alox vs. pH in 0-17cm layer of the soil profile

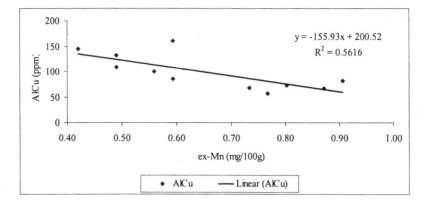

Figure 6. AlCu vs. ex-Mn in 0-17cm layer of the soil profile

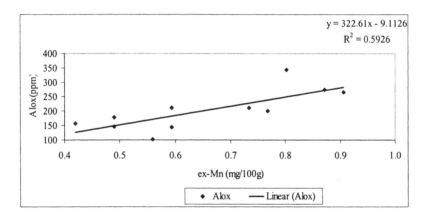

Figure 7. Alox vs. ex-Mn in 0-17cm layer of the soil profile

8.   SPATIAL DISTRIBUTION OF AL SPECIES

Al_{pstot} in 0-17cm and 37-57cm was more abundant in the ridge soils (1.11 ± 0.10%, 1.43 ± .26%, respectively), while the alluvial plains were richer in 17-37cm, and 57-86cm of the soil profile (1.32 ± 0.13%, and 1.36 ± 0.11%, respectively). The concentration of Al_{Cu} was higher in the ridge soils when compared to floodplain soils throughout the soil profile. Both the ridge and floodplain soils exhibited a decreased occurrence of Al_{py} species with depth. Amorphous oxides and hydroxides of Al were more abundant in floodplain soils to depths up to 57cm while the ridge soils were richer in 57-86 cm depth. However, there seemed to be no variation with respect to ridge and floodplain soils in the case of crystalline Al oxides as they were uniformly distributed in all the soils studied. Al_{ox} and Al_{DCB} were the dominant species in the surface layer of floodplain soils with Al_{Cu} being the least abundant; while in ridge soils Al organic complexes were more dominant than Al_{DCB}. The distribution of Al species was more uniform upto 37 cm in floodplain soils unlike the ridge soils.

9.   CONCLUSIONS

The present study on the chemical speciation of aluminum in Delhi soils shows that "pseudo" total Al and Fe are bound predominantly as oxides and silicates throughout the soil profile with only very low percentages in the easily mobilizable pools. 0-17cm and 57-86cm layers of the soil profile are characterized by an abundance of amorphous oxides and oxyhydroxides of Al, while crystalline and "free" oxides of Al, is the dominating fraction in the intermediate layers. Weakly exchangeable and interlay polymers of Al are the least abundant and contribute the least to speciation in the entire soil profile. Al-organic complexes are the least abundant, next only to Al_{Cu} in the profile of the selected soils. The maximum contribution of Al and Fe species to Al_{pstot} and Fe_{pstot} in the soil profile is in the upper layer (0-17 cm), while it is the least in 37-57cm depth. This investigation also emphasizes that phytotoxicity arising from the dissolution and release of ionic Al species is not presently a threat in Delhi soils due to the alkaline soil pH conditions. This is supported by the findings that KCl-exchangeable was not detected in any of the soil samples chosen for the study. The 37-57cm layer needs special investigation with regard to the observation that it has the lowest contribution towards speciation despite being the most abundant layer for Al_{pstot}. As the chemistry of Al, Fe and Mn in the soils is interrelated, future studies could focus on the speciation of these three metals in tandem in the soil profile to provide a better insight into the various forms occurring in the soil environment. More importantly, Al solubility control in these soils is not understood completely. Using the natural spatial heterogeneity of soil solution and the chemistry of the soils in the ridge and the floodplains, batch equilibrium experiments or undisturbed soil column experiments in the laboratory or in situ lysimetry can be conducted to propose the reactive Al phase(s) in these soils that actually controls its solubility.

10.   ACKNOWLEDGEMENT

The financial assistance provided by JNU-UGC for this research is gratefully acknowledged.

REFERENCES

1. Evans, C.E., Kamprath, E. J., Lime response as related to percent Al saturation, solution Al and organic matter content, Soil Sci Soc Am Proc 34, 1970, 893-896.
2. Barnhisel, R., Bertsch, P. M., Aluminum, in: Methods of Soil Analysis, Part 2, Chemical and Microbiological properties, ed. A. L. Page, Agronomy Monograph 9, ASA-SSSA, Madison, WI, 1982, pp. 275-300, ASIN 0891180729.
3. Driscoll, C.T., Aluminum in acidic surface waters: Chemistry, transport and effects, Environ Hlth Persp 63, 1985, 93-104.
4. Driscoll, C. T., Schecher, W. D., The chemistry of aluminum in the environment, Environ Geochem Health 12, 1990, 28-49.
5. Driscoll, C. T., Postek, K. M., The Chemistry of aluminum in surface waters, in: The environmental Chemistry of Aluminum, ed. G. Sposito, 2nd edition, CRC Lewis Publishers, Boca Raton, FL, 1996, pp. 364-418, ISBN 1566700302.
6. Zhu, X. P., Kotowski, M., Pawlowski, L., The relative importance of aluminum solid–phase component in agricultural soils treated with oxalic and sulphuric acids, in: Chemistry for the Protection of the Environment, 3, eds. L.Pawlowski, M.A. Gonzalez, M.R. Dudzinska, W.J.Lacy, Environmental Science Research, vol. 55, Plenum Press, N.York, 1998, pp.245-254, ISBN 0-306-46026-2.
7. Haldar, A., Profiles and speciation of some heavy metals in Delhi soils, Ph. D. Thesis, Jawaharlal Nehru University, New Delhi, India, 1997.
8. Chibbar, R. K., Soils of Delhi and their management, in: Soils of India and their management, eds. B. C. Biswas, D. S. Yadav, S. Maheshwari, The Fertilizer Association of India, 1985, pp. 72-86.
9. Thomas, G. W., Exchangeable cations, in: Methods of Soil Analysis, Part 2, Chemical and Microbiological properties, ed. A. L. Page, Agronomy Monograph 9, ASA-SSSA, Madison, WI, 1982, pp. 159-165. ASIN 0891180729.
10. Juo, A. S. R., Kamprath, E. J., Copper chloride as an extractant for estimating the potentially reactive aluminum pool in acid soils, Soil Sci Soc Am J 43, 1979, 35-38.
11. McKeague, J. A., An evaluation of 0.1 M pyrophosphate and pyrophosphate-dithionite in comparison with oxalate as extractants of the accumulation products in podzols and some other soils, Can J Soil Sci, 47, 1967, 95-99.
12. McKeague, J. A., Day J. H., Dithionite- and oxalate-extractable Fe and Al as aids in differentiating various classes of soils, Can J Soil Sci 46, 1966, 13-22.
13. Holmgren, C. G. S., A rapid citrate dithionite extractable iron procedure, Soil Sci Soc Am Proc, 31, 1967, 210-211.
14. Netush, D., Hopke, P., Analytical aspects of environmental chemistry, John Wiley and Sons, NY, 1983, pp. 219-262.

3

THERMAL TREATMENT TECHNOLOGIES

THERMAL AND BIOLOGICAL DEGRADATION OF SITES CONTAMINATED BY TRANSFORMER OIL

Krystyna Cedzynska

Technical University of Lodz, Institute of General Food Chemistry, 4/10 Stefanowskiego , 90-924 Lodz, Poland Phone + 48 42 6313428; E-mail: krys@snack.p.lodz.pl

Abstract: The destruction of waste transformer oil contaminated with some toxic compounds (e.g., PCB's) was investigated in two ways: biological degradation and thermal treatment.

In both cases, the samples from contaminated sites were rinsed with a solvent to obtain an extract of contaminated transformer oil. The effects of biological degradation were investigated by using a commercial mixture of microorganisms and pure strain under aerobic and anaerobic condition. In the thermal method, a laboratory plasma system was used to decompose the contaminated transformer oil by a direct injection of the oil extract into the plasma system or by melting the extract samples with power plant fly ash in the plasma reactor. For the contaminated transformer oil both methods showed a destruction efficiency of 99.99% and the products of destruction were environmentally friendly.

Key words: Contaminated oil waste, biological degradation, thermal decomposition and plasma treatment, PCB's

1.   INTRODUCTION

A number of capacitors and transformers containing mineral oil with some toxic compounds (e.g., PCB's) are still working or they are stored for future recycling. The unavoidable leaks of oil from transformers during their operation or storage can be released to the environment where they still persist. There are few ways of purifying sites contaminated by transformer oil: biological degradation and chemical or thermal treatment. In all cases, the samples from contaminated sites could be washed with a solvent to obtain an extract of contaminated transformer oil.

Biological decontamination from toxic compounds depends on groups of microorganisms, which use these compounds as the main source of carbon and energy. In our case these groups of microorganisms should degrade not only hydrocarbons from oil but also and maybe first of all oil contaminants e.g., PCB congeners.

The biological methods follow in two ways: aerobic degradation and anaerobic dechlorination [1]. In aerobic process the microorganisms attack congeners via dioxigenase pathway converting the toxic molecules to correspond chlorobenzoic acid, which can be mineralised to carbon dioxide, water and chloride [2,3]. Anaerobic bacteria attack more chlorinated PCB's congeners, through meta- and para- chlorine removal, to obtain less chlorinated ortho- substituted products [4]. In the both cases the products of PCB's biodegradation are environmentally friendly.

A thermal plasma system, to completely decompose the extract of contaminated transformer oil[5], may do the thermal processing. For a complete destruction of highly halogenated hydrocarbons like the PCB's and waste transformer oil, a high processing temperature must be ensured. If these substances are added to conventional incineration plants, problems arise with dioxins and other toxic substances in the exhaust gas. The plasma method with its high process temperatures, the short reaction times and the good process control, the process is an appropriate alternative to treat such material with a low heating value and to convert it into harmless products.

The aim of the work was:
- study the biological destruction process of oil with PCB's using commercially prepared mixtures of bacterial/enzyme complex and some strain.
- study the efficiency of thermal destruction of the extract of the contaminated transformer oil by rotating arc plasma system.
- develop an optimal process performance to find the optimal temperature for the treatment of oil – PCB's in plasma reactor.
- compare the destruction efficiency of biological and thermal methods.

2.   MATERIALS AND METHODS

2.1   Chemicals

The PCB congeners Mix 1 of Supelco were used as the standard. PCBs Mix 1 contains six PCBs congeners in concentration of 10 ppm each: 2,6-

dichlorobiphenyl (2-CB), 2,4,4'-trichlorobiphenyl (3-CB), 2,2',5,5'-tetrachlorobiphenyl (4-CB), 2,2',4,4',5,5'-hexachlorobiphenyl (6-CB), 2,2',3,4,4',5'-hexachlorobiphenyls (6*-CB), 2,2',3,4,4',5,5'-heptachlorobiphenyl (7-CB). The waste transformer oil with PCBs from Paraffine Institute of Warsaw was investigated. The oil consisted in 30% of PCBs congeners.

2.2   Organisms and culture

In the experiment commercial biological compositions of microorganisms cultures especially suited for utilisation of hydrocarbons in polluted soil and water were used for biodegradation of oil and PCBs congeners. These mixtures were Biozyn 300 and 301, NS 20-20 of Bioarcus Company. The pure strain was a PCBs degrading strain of *Pseudomonas pseudoalcaligenes No10086* from German National Collection of Microorganisms (DSMZ).

The condition of the commercial compositions of microorganisms and strain were previously reported [6].

2.3   Chemical analysis

The extraction method with solvent mixture acetone/ n-hexane (1:3) was used to separate the organic solution containing PCBs. T.J.Backer separate disc and Backerbond columns with the special phase were used to extract and purify the PCBs congeners from organic solution. The SRI 8610 Gas Chromatograph equipped with a 30m capillary column DB-5 and ECD detector was used for analysis of oil and PCBs congeners. Identification of peaks and their calibration was made according to Supelco PCBs Congeners Mix. The difference between concentration of oil and PCBs congeners before and after biological process and thermal treatment was found as the level of degradation. Conventional extraction method in the Soxhlet microextractor was used. The solvents for extraction and the chromatographic analysis were selected as follows: cyclohexane, isooctane, benzene and a polar mixture of isooctane with cyclohexane.

2.4   Experimental for thermal treatment

Plasma destruction of transformer oil with PCB absorbed in sites was investigated in two ways of the waste feeding. In the first experiment the sample was „washed up" with a solvent and the fluid extract was then injected into the plasma. The laboratory plasma system with a rotating arc device (ROTARC) has been used as a plasma generator. It was fully described in [7]. High-pressure fluid injector placed on the top of the ROTARC inner electrode produced a highly atomised conical spray. The atomised fluid was injected downstream the plasma jet of the rotating arc reactor. In the second experiment the contaminated extract has been mixed with fly ash from power plant and melted, as it was previously reported [8]. The object of the investigations was the fly ash contaminated with 5,10,20,30% by mass of extract of oil containing PCB's. The fluid extracts have been atomised and injected into the hollow plasma of the reactor chamber. After shock quenching in the alkali scrubber the effluent gas products and also the melted product were analysed.

3.   RESULTS AND DISCUSSION

3.1   Biodegradation

In our study we used commercial microorganisms mixtures especially suited to utilisation of hydrocarbons. The pure strain of *Pseudomonas pseudoalcaligenes* No 10086, one of the most effective PCBs congeners degrades was used in experiment to check the process effects in both (with and without oxygen) conditions in presence of waste transformer oil.

In the cultures with commercial biological mixtures, the most effective biodegradation effects of 2-CB were observed in the solution of Biozyn 301, 52%. The biodegradation of 3-CB was up to 83% in cultures with Biozyn 300 and NS 20-20. In the culture with strain 10086 the biodegradation effects for light chlorinated congeners were over 66%.

The 6-CB and 7-CB biodegradation results in cultures with commercial biological mixtures were from 22 to 37% and from 27% to 47% respectively. The biodegradation effects observed in cultures with strain ·10086 *Pseudomonas pseudoalcaligenes,* were similar to the results obtained for commercial biological mixture (from 18 to 35%). As we previously reported [6], the presence of mineral oil hydrocarbons influences the biodegradation effects of PCB's. In both of the condition, aerobic and anaerobic, the high level of biodegradation is observed for low chlorinated congeners. The results suggest that the microorganisms use PCB's as the source carbon and energy but firstly they degrade mineral oil hydrocarbons which are less resistant to including them into metabolic pathways. The enzymes, which are produced specifically for the hydrocarbon degradation, can not specifically cleavage biphenyl rings of low chlorinated congeners independence of the oxygen presence.

3.2   **Thermal degradation**

In the first experiment the contaminated extracts were obtained and then they were thermally treated by plasma. The analytical results showed that the best solvents for extraction of oil - PCB's were: isooctane, cyclohexane, isopropyl alcohol and their mixture. We have found the average extraction efficiency 95%. The best result 96.2% of the extraction efficiency we obtained for the mixture of isooctane and isopropyl alcohol in the ratio 1 : 1.

The extracts were then atomised and fed into the ROTARC reactor for high temperature treatment. In the first case the atomised extract was mixed with the torch gas (Argon) only. It was a pure pyrolysis, which was effective in the sooting of the reactor walls and it was making the scrubber fluid dirty. The disadvantage of the pure pyrolysis process confirmed our theoretical considerations on thermal destruction of PCB's presented in [9]. To avoid sooting, we fed steam into the reaction chamber in the amount of 10% above the stoichiometry. In this case, which we call the "wet pyrolysis", we obtained the destruction efficiency of oil- PCB's at least 99.99%. The off-gas analysis on the concentration of oil-PCB's were below the detection limit 0.2 ppm .

Beside the experimental activities the constitution of the exhaust gas was studied theoretically using the thermodynamically equilibrium approach. To predict the variety and amount of products appearing during pyrolytic heating of power plant ash and transformer chlorinated oil (PCB) in the plasma furnace the programmes "ChemSage" and "ChemSheet" were used.

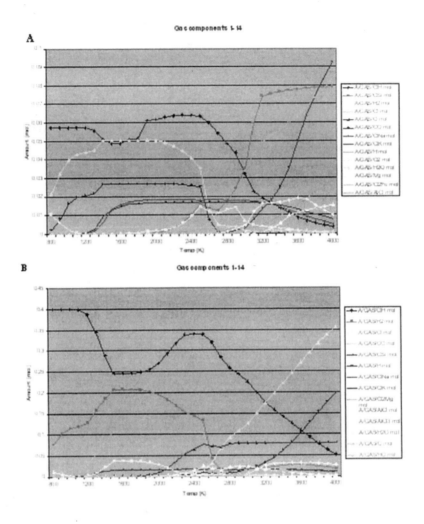

Figure 1. Thermochemistry of power plant fly ash with 5% (A) and 30% (B) of PCB-oil presented for gases only

To check the influence of PCB oil admixture to the fly ash on the thermodynamic conversion of the whole mixture, the calculations were done for various amounts of organic compounds. The results for power plant ash thermodynamic conversion with temperature rise and with oil-PCB's addition are shown in Fig.1 (A and B). These figures show that chlorine appears in the form of HCl with a characteristic content minimum in the temperature range 1300-1700 K. In the range of 1000 –3000 K the off gas is rich in hydrogen. Maximum value of H_2 is determined by the methane decomposition, which occurs above 1200 K. Due to different the decomposition of various oxides and carbon oxidation, the carbon monoxide amount increases significantly in the range of 1600 – 2000 K. Above 2000 K its concentration is at maximum. Carbon monoxide together with hydrogen creates the most caloric off gas in the 2100-2400 K temperature range. The total amount of this off gas depends on the quantity amount of PCB-oil in the fly ash.

Nevertheless, independent of the oil-PCB's mixture added to the power plant ash, the process of plasma pyrolysis should be run between 1400 and 1700 K aiming to obtain the best exhaust gas mixture. Basing on computer simulation in the second experiment, the plasma pyrolysis process was conducted up to the time required for the theoretically found temperatures. During this process, the sooting appeared to be negligible and also less chlorine was found in the off-gas. This is understandable if we consider high temperature in the plasma reactor in which carbon is vitrified and chlorine is partially reduced to salt. The concentration of oil-PCB's in the dry and normal temperature off-gas was also below the detection limit of the GC, which resulted in destruction efficiency above 99.99%. The melted product-slag after cooling was tested on its leaching according to UE Standard [8]. Concentration of chloride and heavy metals ions and PCB's was obtained below the detection limits. The cooled slag was homogenous and vitreous in nature. The final product looks like a having basalt structure and is environmentally acceptable.

4. CONCLUSIONS

Comparing both the methods- thermal and biological treatments, we should decide which of them is better to use at the moment. In the case when the deep layer of sites is polluted by waste transformer oil with PCB's, using thermal methods are the most effective in the destruction of toxic compounds (above 99.99 %). When the pollution is spread over large surface area, the biological methods could be more efficient. However, the specific aerobic microorganisms, screening from the polluted areas, should be used for the high percentage effects of degradation (70-90%). In the thermal method of degradation, the two ways of oil-PCB's feeding (extraction with the extract directly injection and extract mixing with ash to melt and gasified) to the plasma reactor appeared to be of similar destruction efficiency. However, the employing of the melting system has following advantages:
- the system is compact and easy controlled

- the plasma processing can be done as pure pyrolysis - negligible soot is being formed
- steam or oxygen is not necessary to be added to prevent sooting
- very stable basalt-like structure product is obtained.

REFERENCES

1. Abramowicz D.A., "Aerobic and anaerobic biodegradation of PCB: a review", Crit. Rev. Biotechnol. 10, 1990, pp.241-251.
2. Mondello F.J., D.A. Abramowicz, J.R. Rhea, "Natural Restoration of PCB-Contaminate Hudson River Sediments" in Biological Treatment of Hazardous Wastes, (1998), John Wiley & Sons, pp.303-326.

PLASMA TREATMENT OF INORGANIC WASTE

Zbigniew Kolacinski, Krystyna Cedzynska
Technical University of Lodz, 4/10 Stefanowskiego Street, 90-924 Lodz, Poland, E-mail: zbigniew@ck-sg.p.lodz.pl

Abstract: The fully friendly to the environment and the most cost-effective thermal plasma system for waste destruction are here presented. The proposed solutions are directed towards zero emission of any hazardous residuals. This results in no longer waste storage in landfills if the plasma technologies are widely commercialised. The final product (vitrified material) obtained after the plasma treatment is environmentally acceptable. It can return to the environment as an aggregate in the construction industry.

Key words: Thermal plasma, hazardous waste treatment

1.   INTRODUCTION

EU regulations banning the disposal to communal landfill of toxic and hazardous waste are expected to become effective after a year 2002. In the interim period, toxic medical and industrial waste post-combustion residues containing heavy metals, furans and dioxins were being disposed off to landfills throughout the EU with the possibility of leaching into the water table. The above can be resolved by employing of plasma technology for thermal treatment of inorganic and/or hazardous organic wastes.

2.   THERMAL PLASMA POTENTIAL

The thermal plasma is a source of high energy density with temperature of a few thousand degrees and high ultraviolet radiation. These result in: fast reaction rates, high throughput in smaller reactors, heat generation independent of the chemical composition, avoidance of dioxins and furans formation due to oxygen free atmosphere, freedom to select optimum chemistry for the destruction process, melting of high temperature inorganic wastes, easy control, rapid start-up and shut down and flexible treatment process.

The plasma process on an industrial scale is a quite sophisticated and know-how packed technological system. The target applications of the thermal plasma systems presently are:

– treatment of the most toxic and hazardous industrial wastes,
– clean fuel generation from organic hazardous wastes,
– vitrification of solid leftovers from conventional incineration plants and nuclear power stations,
– vitrification of contaminated soil.

The applications specified above have been demonstrated elsewhere [1,2,3] on the base of some research results obtained in the Plasma Technologies & Environmental Protection Group (PTEP) of the Lodz Technical University. One of the most important applications is the plasma vitrification combined with waste incineration. This gives no post-incineration residues and makes the system friendly to the environment. Two general technical solutions of this approach are here considered.

3.   PLASMA TOWARDS ZERO WASTE RESIDUALS

3.1   Present day problem

EU regulations banning the disposal to landfill of toxic and hazardous wastes are expected to become effective in year 2002. In the interim period, toxic medical and industrial waste post-combustion residues containing heavy metals, furans and dioxins are being disposed off to landfills throughout the EU with the possibility of leaching into the water table being ever present. In addition, other hazardous fluid/solid wastes, notably contaminated oiled wastes, pesticides and used chemicals are placed in supervised storage, presenting increasing disposal problems for future generations. The latest available EU statistics [4] show that approximately 10,302,000 tonnes of hazardous waste, of which 1,030,200 tonnes was oil waste, was produced in 1990 within the Community. Similar problems arise in Central and Eastern European Countries. Lack of regulations causes the disposal of highly toxic waste in the ordinary communal landfills, which are not protected for leaching. Special problem arises with the polluted military sites and airfields, which are saturated with toxic oily chemicals. The large stream of hazardous residues comes from waste incinerators as a result of the following shortcomings of the fuel burning process:

- The thermal treatment of some wastes (e.g. industrial, medical, and military) through their incineration results in the formation of relatively highly toxic residues. The ash residue being a secondary waste is sometimes more toxic than the primary solid feed.
- Waste incineration leads to a decrease of the waste mass only. From 15 to 30 % of feed mass is still the ash residue.
- Dioxins, furans and heavy metals can be released as air borne pollutants in the ash, which needs to be landfilled with special care (i.e. mixed with cement) due to a danger of ground water contamination.

The above can be resolved by employing of plasma technology for thermal treatment of inorganic and/or hazardous organic wastes.

3.2   Plasma treatment strategies

A major advantage of plasma processing is that the heat input may be accomplished in an atmosphere of any desired composition and reactivity. In practice there are only a few variations of chemical strategies available for thermal processing i.e. pyrolysis, oxidation, reactions with hydrogen and water. They were already reported elsewhere [5]. The most cost effective and friendly to the environment are the approaches of plasma employing for zero-waste fuel generation or for zero-waste incineration.

3.3   Zero Waste Fuel Generation

Thermal plasma is ideally suited for the massless heating of solid and liquid hydrocarbons to convert them into by-products having recoverable values. The gas mixture contains primarily hydrogen and carbon oxide, approximating reformed natural gas, but the molten slag-metal mixture solidifies into a non-leachable solid. The products rich in hydrogen and carbon oxide may be used in the manufacture of methanol, or used as fuel to generate steam for electric power or directly run the gas motor or turbine to generate electricity.

Gas fuel generation from wastes is very attractive and environmentally friendly activity. Several computer analysis of the solid waste conversion into gas fuel was done in the PTEP Group [6].

As an example we present here some computation results for the plasma treatment of the hospital patients room waste with advantage of plastics, cellulose and non-organic metal and glass parts. To check the efficiency of gas fuel generation from this waste we have simulated the limited oxidation within temperature range 1000 - 2500 ^{0}C. We have assumed that in 1kg of dry waste are (by weight) 60% of plastics, 15% of silicon glass, and 10% of calcium glass, 10 % of papers and 5% of metals. Moreover 15% of moisture content was assumed in the calculations. Considering the oxidation process it can be seen in Table 1 that the oxidation rate resulting in CO_{max} production may be achieved in relatively small amount of oxygen supplied (approximately 0.6 kg) keeping the same calorific value of the product gas. The best temperature range for the plasma processing is from 1500 to 2000 ^0C, because in this range the glass fraction melts and encapsulates even refractory metals. In such process conditions the net energy rate Table 2. required for plasma torch is below 1 kWh/kg, but about the same value of energy is stored in energy. There are two main advantages of fuel gas production with plasma compared to incineration in the oxygen starvation conditions: the highest possible calorific value, the process temperature enables ash and slag melting (vitrification).

Table 1. Products of plasma limited oxidation

Temperature	Oxygen at CO_{max}	CO	H_2	H_2O	CO_2	C
^0C	kg		Mass percentage			
			Volume percentage			
1000	0,56	76,0	4,0	4,9	3,9	11,0
		72,767	22,991	3,196	1,046	
1500	0,69	84,5	7,7	3,8	2,8	1,0
		63,548	33,845	2,004	0,614	
2000	0,60	86,8	8,0	2,0	0,2	1,0
		63,443	35,100	0,974	0,044	

Table 2. Energy balance at plasma limited oxidation

Temperature	Oxygen at CO_{max}	Net energy consumption	Calorific value	Energy in gas fuel
^0C	kg	kWh	MJ/m^3	MJ kWh
1000	0,56	-0,50	12,662	2,363 0,656
1500	0,69	-0,35	12,672	3,063 0,851
1500	0,69	-0,35	12,672	3,063 0,851
2000	0,60	+0,90	12,673	3,188 0,886

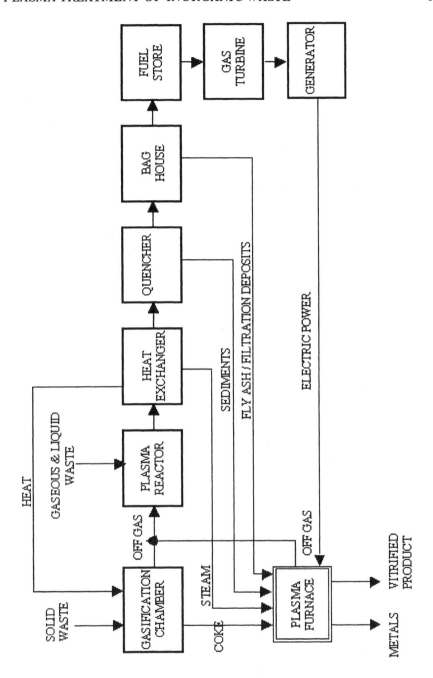

Figure 1. Flowchart of plasma zero-waste fuel generation process

In Figure 1 the flow chart of gas fuel generation together with vitrification of all solid residues is presented. The system is self-supplied in the electric power and it employs one plasma furnace for vitrification of solids and producing fuel gas from the coke and one plasma reactor for gaseous and liquid waste destruction. The excess of energy in the form of electricity or gas fuel can be sold.

4.   ZERO WASTE INCINERATION

The incineration of certain wastes (industrial, medical, and military) results in the formation of relatively highly toxic residues. The toxic leftovers (ash, slag, filter deposits, sedimentation residues) could be easily disposed in landfills assuming that they were first immobilised and converted into a non-leaching product. When the leftovers are heated to a sufficient temperature, their elements, including minerals and toxic heavy metals, melt and glassify. Even partial solidification (vitrification) of those residues requires the temperature above 1700K, which is not available in most incinerators but easily reachable in thermal plasma reactors. Temperatures of the order of 10000 K are typical for arc in plasma furnaces and all inorganic residues can be solidified. The system of plasma vitrification of ash produces a chemically stable and mechanically resistant product. After vitrification the mineral product looks like glassy, basalt structured lava (even of higher than basalt mechanical strength), and its main components are silicon, aluminium and calcium oxides in the form of chemically inactive compounds, resistant to flushing.

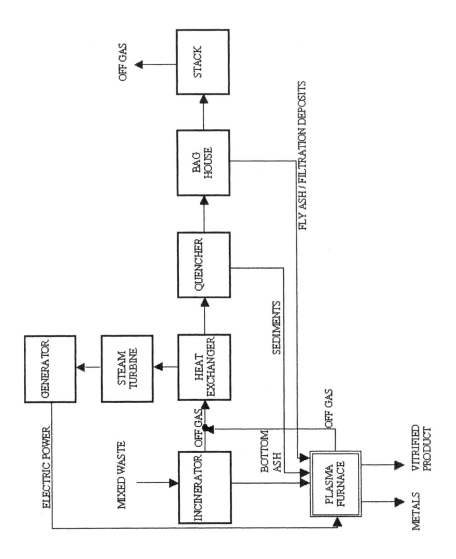

Figure 2. Flowchart of plasma zero-waste incineration process

The DC plasma furnace was designed in the Technical University of Lodz and applied to treat the ash samples with high degree of flexibility [7]. The process of plasma vitrification of ash and slag ends in a chemically stable (Table 3) and mechanically resistant product [8,9], which is safe to store or it can be used as constructional material even in the form of bricks or roof tiles.

The plasma vitrification can be used also for asbestos waste which is a hazardous, fibrous, dusty material causing danger for human beings such as lung cancer – mesothelioma.

In Figure 2 the flow chart of classic fuel burning incinerator combined with plasma vitrification system of all solid residues is presented. The system is self-supplied in the electric power and it employs only single plasma furnace for vitrification of solids. This way every classic waste incineration plant can be converted to zero waste emission system. The excess of energy in the form of electricity or steam can be sold.

5.   CONCLUSIONS

Worldwide, there are numerous plasma system designs for treatment of all types of wastes. Economical considerations limit their commercial applications to the most profitable actions. Presently they commercially operate: in Switzerland and Germany for low level nuclear waste vitrification, in France and the USA for asbestos waste vitrification, in the USA and Australia for hazardous waste treatment, in Japan and France for municipal fly ash vitrification. The most of installations is working in Japan because there 70% of municipal waste is incinerated and the ash can not be used as landfill. EU Regulations banning the disposal to landfill of toxic and hazardous wastes after year 2002 may cause wider use of plasma waste destruction technology in Europe.

REFERENCES

1. Kolacinski, Z. "Prospective of Thermal Plasma Application to Waste Treatment" - Invited lecture for the VIII Int. Symp. SAP&ETEP 97, Lodz, September 1997, Post-Conf. Materials, 1998, pp. 117-126.
2. Cedzynska, K., Kolacinski, Z. "Conversion of Waste Incinerators into Environmental Friendly Plants", ANQUE, Puerto de la Cruz, 1999.
3. Campbell, L.C., Kolacinski, Z., Stewart, M., Dokimuk, J., "Spectroscopic Measurements of Plasma Flame Temperature, ", Proceedings of the 11th Int. Conference on "Plasma Chemistry and Plasma Processing" Loughborough, Leicestershire, England, Vol.2, 775 - 781, 1993.
4. Environment Statistics Yearbook Eurostat, 1996.
5. Kolacinski, Z., Cedzynska, K., "Chlorinated Waste Destruction by Plasma Pyrolysis", Progress in Plasma Processing of Materials, New York, Wallingford, 1997, p. 493-500.

4

ENVIRONMENTAL ASPECTS OF AGRICULTURAL PRACTICES

FORMER U.S. DEPARTMENT OF AGRICULTURE/COMMODITY CREDIT CORPORATION GRAIN BIN PROJECT

Jeffrey L. Field
Drinking Water / Groundwater Management Branch, U.S. Environmental Protection Agency, Region 7, 901 N 5th Street, Kansas City, KS 66101: Phone: (913) 551-7548

1.   INTRODUCTION

The U. S. Environmental Protection Agency, Region 7 has been actively working with state agencies within Region 7 to conduct sampling activities at former U.S. Department of Agriculture/Commodity Credit Corporation (USDA/CCC) grain storage facilities located in Iowa, Kansas, Missouri, and Iowa. The intent of this project was is to protect public health and drinking water supplies by addressing possible contamination of groundwater by carbon tetrachloride, a component of a grain fumigant used at these grain storage facilities. Sampling activities have been completed at those facilities identified with the assistance of USDA/CCC. These activities have revealed numerous detections of carbon tetrachloride above and below the federal drinking water standard of five micrograms per liter.

2.   DISCUSSION

The Commodity Credit Corporation (CCC) initiated a large-scale grain bin construction program in the late 1940's for the purpose of storing surplus grain. These temporary facilities were used intermittently in areas where commercial storage was unavailable. At its peak during the 1950's, CCC operated grain storage facilities on leased property at several thousand locations nationwide. Some of the grain was stored for a period of several years before being sold. During the storage period it was sometimes necessary to fumigate the grain to control destructive insects. The most commonly used fumigant during this period was an 80/20 mixture of carbon tetrachloride (CCl_4) and carbon disulfide (CS_2) respectively. Ethylene dibromide was also present in these fumigants, but in smaller amounts. The liquid product was applied directly onto the grain from the top of the storage bin. Fumigation using CCl_4 combinations was accepted grain storage practice for insect control by the industry at the time and was not replaced with newer contact insecticides until the early 1960's. CCC substituted the insecticide Malathion for CCl_4 in the early 1960's. Also in the early 1960's, CCC began selling off many of its storage structure to farmers who used them for farm storage. CCC terminated its grain storage program by the early 1970's and sold all existing grain storage bins and equipment. CCC records relating to storage and fumigation of grain stored in the bins, and their disposal, were not retained following termination of the program in the early1970's.

Ground water contamination by carbon tetrachloride was first discovered in Region 7 at Waverly, Nebraska, in 1982 during a routine test of the Public Water Supply (PWS) by the Environmental Protection Agency (EPA). Initial contaminant levels indicated 200 parts per billion (ppb). Subsequent sampling in 1986 revealed concentrations up to 3128 ppb. In 1986 EPA established a recommended maximum contaminant level (RMCL) in ground water for CCl_4, and in 1986 they established the maximum contaminant level (MCL) for CCl_4 at five ppb. By 1986, concerns were raised about the past fumigation practices by CCC at Waverly, and other CCC grain storage locations in Region 7. At Waverly, EPA began conducting tests to locate the source of contamination. It was determined that the former CCC grain storage site was the source for the contamination.

Numerous sampling activities have been conducted in Iowa, Kansas, Nebraska, and Missouri since the initial detection of carbon tetrachloride at Waverly.

3.   RESULTS

3.1   Iowa

The EPA has sampled 175 former USDA/CCC grain storage facilities. Only three locations had detections of carbon tetrachloride. None were over the drinking water standard.

3.2   Kansas

The Kansas Department of Health and Environment and the EPA has sampled at 273 former USDA/CCC grain storage facilities. These efforts revealed 38 locations (14%) with detections of carbon tetrachloride, with 22 (8%) of these locations having levels exceeding the established drinking water standard of five micrograms per liter.

3.3   Missouri

The Missouri Department of Natural Resources and the EPA have sampled 77 former USDA/CCC grain storage facilities. Nine locations (12%) had detections of carbon tetrachloride, with four (5%) of these locations exceeding the drinking water standard of five micro grams per liter.

3.4   Nebraska

The Nebraska Department of Environmental Quality, Nebraska Department of Health and Human Services, and the EPA has sampled 308 former USDA/CCC grain storage facilities. These efforts revealed 80 locations (26%) had detections of carbon tetrachloride, with 35 (11%) of these locations exceeding the drinking water standard of five micro grams per liter.

As a result of these sampling efforts two locations (Bruno, Nebraska and Waverly, Nebraska were placed on EPA's National Priorities List (NPL) for clean-up of carbon tetrachloride contamination. The EPA entered into an Administrative Order on Consent with USDA for the groundwater contamination at Murdock, Nebraska. All locations that had detections that exceeded the drinking water standard were provided alternate sources of drinking water (i.e., bottled water, connection to public water supply).

4.   CONCLUSION

Carbon tetrachloride has not been used as a grain fumigant for nearly 16 years, yet it has recently been detected in soils and groundwater at many locations in Region VII. This project has conducted sampling activities at 829 locations in the four states that make up this region. A total of 130 locations (16%) had detections of carbon tetrachloride in water samples, with 61 (47%) of these with levels exceeding the drinking water standard of five micro grams per liter. Based on the data that has been collected, it is evident that past fumigation practices at former USDA/CCC grain storage facilities has negatively impacted some ground water sources in Region VII.

AGRICULTURAL BIOTECHNOLOGY: PUBLIC ACCEPTANCE, REGULATION AND INTERNATIONAL CONSENSUS

Nathan Richmond
Masters Candidate in Public Affairs, Washington State University, E-mail: RichmondNJ@hotmail.com

Abstract: In this paper, I will analyze the relationship between market integration in the international food system and three interrelated debates in the biotechnology sphere: public acceptance, regulation and international consensus. I will examine the construction of EU public opinion, its impact on EU policy and its indirect effects in the global food system including the EU-US regulatory conflict. I will also investigate the extent to which EU market forces have forced other nations to respond, in some cases leading to international consensus. In closing, I emphasize the importance of market integration in driving countries to respond to public opinion in the EU.

1. INTRODUCTION

The issue of agricultural and food (agri-food) biotechnology is a highly politicized global debate. The European Union (EU), which is the largest agricultural importer, is the foremost advocate of caution while the United States (US), which is the largest agricultural exporter, is the leading supporter of biotechnology. Their debate yields insight into how nations and trade blocs are economically and politically interdependent in the global food system. The system is bound and governed according to international agreements, which set the rules of engagement between importing and exporting countries, producers and consumers. This paper will argue that three key interrelated debates in agri-food biotechnology can be explained through global market integration: public acceptance, regulation and international consensus.

This paper will begin by briefly defining the term biotechnology. The second part provides evidence to support the underlying assumption in this paper that agri-food markets are globally integrated through trade and are significantly impacted by biotechnology. Section three examines the relationship between public acceptance, consumer preferences and policy in the EU. It will also consider the forces shaping public acceptance. Section four, explores the points of regulatory contention between the EU and the US. Section five, investigates the role of economic interests in forming international consensus. Specifically, it will analyze how changes in public opinion influence the economic interests of others in the food system. These changed interests correspond with shifts toward international consensus. The final section will summarize and discuss the arguments covered in this paper.

2.   BIOTECHNOLOGY

Biotechnology refers to a broad set of traditional and modern scientific techniques that change life forms (organisms) to provide benefits to human culture. However, in this paper, biotechnology will refer to the new DNA, molecular and reproductive techniques, encompassing terms such as Genetically Modified (GM), Genetically Modified Organism (GMO) and Genetic Engineering (GE).

3.   MARKET INTEGRATION AND
BIOTECHNOLOGY

It is widely held that the agri-food system is globalized (Murdock, Marsden, & Banks, 2000, p. 109). Mardsen and Arce (1995, p. 1264) describe global market integration as "transnationalization" where production and consumption transcend national borders and regulations. This sentiment is echoed in the final report of the EU-U.S. Biotechnology Consultative forum (2000, p. 6), which described the spread of inter-linking economic and technological developments in the global system. Trade statistics indicate that the world trade in agriculture was worth $558 billion while the trade in food was $442 billion in 2001 (World Trade Organization, 2001). Together these sectors accounted for 16% of the world merchandise trade.

According to a report by the International Service for the Acquisition of Agro-biotech Applications (James, 2001) the global area of biotechnology crops grew from 1.7 million hectares in 1996 to 52.6 million in 2001. Importantly, the report also found that four countries accounted for 99% of the total area used to grow biotechnology crops. These four countries also fell among the leading 15 agricultural exporters, including the US, which grew two-thirds of the world's biotechnology crops. These figures support this paper's main assumption that agri-food markets have integrated globally and that growers in some countries have rapidly adopted biotechnology crops.

4.   PUBLIC ACCEPTANCE

After approving its first biotechnology crops in 1996, the EU would learn through a series of public mobilizations and consumer boycotts that public opinion was opposed to agri-food biotechnology.   These grassroots movements impacted voting, consumption, and media behaviors, which prompted the EU to become more responsive to the demands of its citizens (Haniotis, 2000, p. 85).   Kalaitzandonakes (2000b, 75) cites, Moynagh (2000), Sylvander, B., & le Floc'h-Wadel (2000) and Beranger (2000) in arguing that consumer opinion has "invited" movements in European policy. Moynagh, (2000, 114) for example, argues that animal welfare policy is not an act of trade protectionism but a reaction to significant consumer opinion. Kleter et al. (2001, p. 1105) explains EU-US regulatory differences through variations in public opinion.   Regardless, of the EU's policy rationale, it is evident from the Eurbarometer surveys carried out in 1996 and 1999 that Europeans are increasingly opposed to agri-food biotechnology (Gaskell et al., 2000, 938).

However, the debate has not been over whether or not the public is opposed to biotechnology. Instead, it focuses on the role that public opinion should have in the formulation of public policy (EU-U.S. Biotechnology Consultative Forum, 2000, pp. 6-7). The US State Department argues that "European policy-makers should not see themselves as prisoners of public opinion" but rather should win public trust through leadership, basing its decisions in science (Larson, 2002b).   The US position is based on the premise that science should be the basis for risk assessment and trade. Ironically, some have argued that the US will not initiate a legal claim under the World Trade Organization (WTO) because of the potential for a consumer backlash in the EU and US (Pollack & Shaffer, 2000, p. 49). Fears of a backlash are supported by evidence that national biotechnology debates have synchronized across national borders through the media, though the US and European debates have remained relatively independent (Bauer et al., 2000). The US position implies that public opposition is rooted in fears over safety so that reinvigorating the role of science in decision-making will assuage fears.

5.   SOCIAL CONSTRUCTION OF VALUES AND RISKS

However, some argue that public opposition in the EU is based on more than scientific factors. Beck argues that experts and science are unable to determine risk because it entails the calculation of cultural factors (Beck, 1999). Gaskell (2001, p. 111) argues that societies evaluate costs and values differently and, as a result, may make very different choices based on the same scientific evidence. Gaskell (2001, p. 107), for example, cites variations in optimism toward technology as an explanation of EU-US regulatory divergence. Levidow (1999, p. 20) argues that risks are culturally selected and emphasized so that no one scientific basis is applicable in both cultures. Perhaps, science is a legitimizing force in what Beck (1999) described as a struggle to define the issue and, in so doing, it's political outcome. Bellows (1997) argues that an issue will be resolved according to the consumer's image of a biotechnology application. Kauppinen (2002, p. 182) argues that consumers value moral and political aspects of a product. According to Forsman and Paananen (2001, p. 1), consumers value food according to its impact on the environment, society and economy. These valuations are made possible through the consumer's linking of agriculture and food with images of the environment and nature (Gaskell, 2001, p. 111; Levidow, 1999, p. 20; Murdock, Marsden & Banks, 2000, p. 108). Ermann (2002, p. 7) argues that food can be valued based on its spatial associations to culture, landscape, traditional farming techniques and heritage. It is likely that the success of farmers' markets in Europe have benefited from these consumer values (La Trobe & Acott, 2000). This view of European culture contrasts with the US where agriculture is viewed as an industry and exclusive of the natural environment (Levidow, 1999, p. 20).

6.   IMPORTANCE OF PUBLIC ACCEPTANCE

If perceptions of biotechnology are socially constructed then public opinion would certainly constitute what Gaskell et al. (2000, p. 938) termed the "second hurdle" for regulators, industrialists and political supporters of biotechnology. The importance of this "hurdle" is heightened by the fact that legislation in the EU is created by EU Member State governments rather than by central agencies, as is the case in the US. It is likely that the former will account for "social factors" more so than a central agency that is relatively insulated from public pressure (Pollack & Shaffer, 2000, 44). The importance of public acceptance centers on its indirect impact on the global food system through EU institutions.

7.   REGULATION

Conflicts over regulation arise when differences have adverse economic affects on others in the food system. This is true in the case of the EU-US trade conflict over biotechnology where conflicts stem from important economic stakes (Roberts, 2000). While both the EU and the US agree that it is appropriate for governments to protect the safety of food and agriculture produced from outside their borders, these measures can also be employed as non-tariff barriers disguised to protect domestic markets. Because biotechnology is a strategic technology for the US's performance in world markets, the EU regulations will significantly impact the degree to which the US will be able to realize the technology's potential (Kalaitzandonakes, 2000a).

8.   EU-US DIFFERENCES

Most of the US's problems stem from the EU's de factor moratorium on biotechnology approvals, which has been in place since 1998. According to one French source, there are no signs that the moratorium will be lifted until traceability and labeling laws are in place (Agra Europe, 2002, pp. EP 6-7). US undersecretary of State, Alan Larson echoes the importance of this issue as a sticking point between the EU and the US (Larson 2002b).

8.1   Labeling

The EU requires positive labeling, i.e. "may contain GMOs", for biotechnology products and products derived from biotechnology. The US, on the other hand, argues for voluntary negative labeling, i.e. "non-GMO", because positive labeling would have adverse and unjustified effects on trade. The US Soybean association, for example, claims that the cost of complying with the requirement may cost $4 billion in trade (Larson, 2002b). The EU bolsters its position as a "consumer's right to know" (Haniotis, 2000, p. 85).

Also at issue, is the EU's threshold for adventitious or accidental contamination. The EU Parliament recently lowered the Commission's proposed 1% threshold to .5%, which the US claims will add significant costs, be difficult to comply with and effectively restrict trade (Larson, 2002b). The EU proposal would also require labeling for highly processed products that do not have a detectable presence of biotechnology (Larson, 2002b).

8.2   Risk assessment

A second area of contention pertains to the process of approving biotechnology products. The US argues that decisions should be based on science and insulated from political pressure, while the EU accounts for "social factors" as well as science in their decision-making (Pollack & Shaffer, 2000, p. 44).   The EU formalized this approach under the precautionary principle. It can be invoked in situations where scientific uncertainty may be low but social uncertainty is high, causing conflicts when adverse economic are at stake (Kalaitzandonakes, 2000b, p. 76). Another difference centers on the scope of the assessment, which in the EU extends to the entire process of production (Haniotis, 2000, p. 86). The US adapted an existing regulatory framework to cover agri-food biotechnology so those end products are evaluated on the basis of safety. Conflict erupted when the EU invoked a precautionary ban on biotechnology products deemed scientifically safe by US authorities. In this way, control over EU policy is a contest between external trade forces and internal political pressures. The economic stakes are sufficiently high that a recent report by the Cato Institute's Center for Trade Policy urged US negotiators to insist on removing democratic politics from approval processes, relying instead on wholly scientific standards (Bailey, 2002, p. 14). The recommendation in this report implicitly recognizes that the EU's policies have important economic consequences for the US and that developing a helpful consensus is in the US's economic interests.

9.   INTERNATIONAL CONSENSUS

If public opinion shapes EU policy, then a certain amount of uncertainty surrounds the future of agri-food biotechnology. The lack of equilibrium between the EU and the US adds to the global uncertainty over the issue since both are powerful political and trade powers and have yet to find common ground.   This section will interpret the consequences of this uncertainty as the national and industrial responses to changes in their economic interests.

10.   ECONOMIC INTERESTS

Caulkin (1999) argues that European activist groups altered the markets for agri-food biotechnology by appealing to consumers. Arce and Marsden (1993, p. 303) describe a new type of consumerism where information about the environmental and social implications of products forces producers to respond to their political demands as expressed through the market. Consumer demand, therefore, externally shapes the behavior of producers and suppliers in other countries as well as in their own (Mardsen & Arce, 1995, p. 1267). Participants in a Food and Agriculture Organization (FAO) forum (2002, section 2.2) recognized the economic impacts of gene flow from biotechnology crops to non-biotechnology crops. They argued that contamination could threaten the farmer's capability to supply specialty markets, which is especially important for developing nations whose economies rely on the export of a small number of agricultural products.

Nations may also view these regulatory and market changes as economic opportunities. For example, consumers are willing to pay three times more for open roost chickens than their traditionally grown ones (Moynagh, 2000, pp. 113-114).   Kleter et al. (2001, p. 1105) argues that the initial biotechnology applications introduced in the EU had very little benefit for the consumer and suggests that technologies oriented toward the consumer should be advanced instead. Brookes (2002, pp. 14-15) argues that there will be key shifts in the biotechnology and non-biotechnology derived soybean markets. He predicts that demand for non-biotechnology products will rise in Asia and sharply in the EU while global supply will remain static or even decline. As demand for non-biotechnology soybeans could outstrip supply, the resulting financial incentives would lead farmers to plant non-biotechnology varieties (Brookes, 2002, p. 16).

Organic markets are also an indication of the market opportunities for non-biotechnology crops. Sylvander and le Floc'h-Wadel (2000, p. 97) argue that organic markets in Europe are driven by consumer demand for products that are compatible with the environment and better animal welfare practices. Hau and Joaris (2002) define organic farming as an approach to agriculture that is socially and environmentally sustainable, which implies that its production is oriented toward consumer values. The success of ogranics is demonstrated by reports that its area of production in the EU has more than tripled between 1993-1998 (Foster & Lampkin, 2000, p. 4). An FAO report (2001, Introduction) covering Austria, Belgium, Denmark, France, Germany, Italy, Japan, the Netherlands, Sweden, Switzerland, United Kingdom, and US found that sales of organic fruits and vegetables increased by 20 to 30 percent in most countries in the latter part of the 1990s. Some of these growth rates were much higher after the BSE crisis, including Italy where after the first detection of BSE in 1998 organic fruit and vegetable retail sales grew at a rate of about 85%.

10.1   National responses

External regulations can shape the economic interests and behaviors of powerful countries with large domestic markets too. The US, for example, argues that the EU's labeling requirement could threaten 4$ billion in US biotech soybean and meal exports and claims that US farmers may choose to plant non-gm corn in order to sell in world markets (Larson, 2002a). The anticipated US response to these adverse economic impacts has been legal action under the WTO. On the other hand, we have seen other strategies to develop consensus through harmonization with the market.

On March 20, 2002, the People's Republic of China (PRC) Ministry of Agriculture implemented its biosafety rules (Wang, 2002). Despite the fact that the PRC has the largest plant biotechnology capacity outside of North America, the biosafety rules are very similar to that of the EU. Their announcement was also "coincidentally" made at a time when the US was considering legal action against the EU (Qin, 2002). While some argue that the framework is a response to WTO fears that small-scale farmers will not be able to compete with imports, it is also argued that their own exports may suffer if it's not perceived as having a strict agri-food biotechnology policy (Kahn, 2002). China has also harmonized temporarily with the US by extending interim provisions to protect $1 billion in US soybean imports (AgraFood Biotech, 2002b, p. 17).

Brazil's possible new administration has indicated that it will likely continue its ban on GMOs, citing premium markets that its competitors in the US and Argentina cannot reach because they plant biotechnology varieties (AgraFood Biotech, 2002a, p. 8). Aware that the EU will be testing imports for biotechnology content, Brazil has voiced its intention to enforce the ban on biotechnology crops and pre-test agricultural products destined for the EU.

Zambia's decision to refuse biotechnology grain aid in spite of famine conditions facing the region was intended to prevent seeds from cross-pollinating with other crops and thus endangering agricultural exports to Europe (Carroll, 2002). Clearly, in the Zambian case economic interests in EU markets overrode domestic pressures to accept the food aid. Mozambique, Malawi, Zimbabwe and Lesotho, which were also facing food shortages, only accepted biotechnology grain that had been milled into flour to prevent the planting of biotechnology seeds.

10.2   Industrial responses

Industry too has contributed toward the establishment of an international consensus. According to Mardsen and Arce (1995,1274) the retail and corporate sector is often the forerunner in the regulation of food safety and quality. In this way, public opinion and consumer choice give large retailers great power in setting standards for their suppliers (Murdock, Marsden & Banks, 2000, p. 109).

European food retailers proceeded with caution and in concert with public opinion, activist organizations and the media, avoiding outcomes contrary to their economic interests (Pearce & Hansson, 2000, p. 457). This example confirms that a product's value is constructed at the processing, distribution and retailing levels (Arce & Marsden, 1993, pp. 293-4). Seed companies too have been faced with economic liabilities arising from the discovery of biotechnology in corn crops, where fields were destroyed and companies were required to compensate farmers for their losses (Nature Biotechnology, 2002). Likewise, organic growers in Canada have pursed legal action against seed companies after unintended contamination disqualified them from organic certification (Nature Biotechnology, 2002).

More recently the Grocery Manufacturers Association (GMA), which represents major food processors, has expressed concern that pharmaceutical producing corn could lead to contamination of foods and heighten consumer fears (BioCentury, 2002, p. A15). The GMA is reacting to the failure of a biotechnology company to prevent the possibility of contaminating food crops with ones designed to produce pharmaceuticals. The Biotechnology

Industry Organization reacted by declaring a moratorium on the growing of pharmaceutical-producing corn in the US Corn Belt and canola in the Canadian canola belt because there are currently no completely reliable techniques for preventing cross-pollination. The voluntary ban applies to areas where it is of considerable economic importance to prevent the spread of biotechnology crops into ones likely to be used for food or animal feed (AgraFood Biotech, 2002c, p. 21). These examples reinforce the relationship between economic interests and the establishment of consensus.

11.   PATHS TO INTERNATIONAL CONSENSUS

Gaskell et al. (2000, p. 938) argues that consensus on agri-food biotechnology should comprise both scientific as well as moral and ethical dimensions underlying public concerns. This argument recognizes that science becomes political when it comes into conflict with public values and opinion (Sundloff, 2000, 137). Decisions in this context become political and thus their resolution concerns not only science but also the establishment of common agreement (Sundloff, 2000, p. 137). This is not too far from a finding in the final report of the EU-U.S. Biotechnology Consultative Forum (2000, p. 22), which prescribed a transatlantic dialogue to examine the differences, especially where non-scientific issues are likely arise. So far, however, the EU and US seem unable to find sufficient common ground, leading many observers to ponder the likelihood of a legal action under the WTO.

11.1   Cartagena Biosafety Protocol

On the contrary, the Cartagena Biosafety Protocol (CBP) is an example of how countries have established common ground. So far, 25 nations of the 50 required have ratified the CBP in addition to the 103 that have signed, indicating their support and plans to join it in the future (Convention on Biological Diversity, 2002). While the US is unable to join the CBP, it was allowed to participate in its negotiation, during which two main issues divided the US from the EU (Pollack & Shaffer, 2000, 52). The first was its precautionary clause, which like the precautionary principle says that lack of certainty can justify blocking an import. It also allows for the consideration of certain socio-economic factors. The second point of contention was over a clause, which states that the protocol will not be subordinate to other international agreements, i.e. the WTO. The CPB enjoys relatively broad-based international support and is a step toward establishing consensus, which would arguably favor the EU position.

11.2   Legal challenge

The US has indicated that it is seriously considering a legal resolution under the WTO in order to establish a science-based consensus between the EU and US (AgraFood Biotech, 2002b, p. 17; Larson, 2002b). However, some predict that a trade war is less likely than an eventual discussion and compromise (Pollack & Shaffer, 2000, p. 53). Some reasons cited include the experience of the US's 1995 to challenge to the EU's ban on US hormone treated beef. The US won on appeal but the EU refused to permit imports of beef after organized public mobilizations against the WTO ruling. There are also predictions that a WTO proceeding could cause a consumer backlash in the EU and the US. These backlashes represent potential shifts in public acceptance, which would have negative impacts on US economic interests.

12.   CONCLUSION

This final section will sum up the major arguments covered in this paper. First, it was established that global agri-food markets are significantly integrated through trade. Secondly, public acceptance proved to be a key factor in the formulation of and justification for EU trade policy. While the US urged the EU to base its decisions on science, it is not clear that the EU is ready to sidestep public opinion. There is evidence to suggest that EU public acceptance is factors environmental and cultural implications associated with the creation of the product. Third, regulatory conflicts erupt when differences have adverse impacts on trade. This is true in the case of the EU-US conflict over biotechnology where EU regulations impinge upon the US's significant economic stake in biotechnology. However, public opinion is lodged behind the main points of contention so that the US is essentially struggling against the public for control over EU policy. Fourth, this paper has reviewed evidence that international consensus may be reached through market forces. Public opinion has impacted EU institutions and indirectly reshuffled the economic interests in the global food system. Finally, it is important to recognize that market integration in the global food system has forced industrial and governmental decision-makers to respond to public opinion in the EU.

REFERENCES

1. Agra Europe. (2002). EU governments in GMO quagmire. October 18, pp. EP 6-7. Agra Europe, London.
2. AgraFood Biotech. (2002a). Tough GM decisions for new Brazil administration. October 29, No. 92, p. 8. Agra Europe, London
3. AgraFood Biotech. (2002b). WTO challenge to EU biotech policy 'seriously' contemplated'. October 29, No. 92, p. 17. Agra Europe, London
4. AgraFood Biotech. (2002c). Some GM crops kept clear of food fields. October 29, No. 92, p 21. Agra Europe, London
5. Arce, A., & Marsden, T. K. (1993). The Social Construction of International Food: A New Research Agenda. Economic Geography, Vol. 69, No. 3, pp. 293-311.

5

CHARACTIZATION AND QUANTIFICATION OF CHEMICAL CONSTITUENTS

RAPID ON-SITE ENVIRONMENTAL SAMPLING AND ANALYSIS OF PROPELLANT STABILIZERS AND THEIR DECOMPOSITION PRODUCTS BY PORTABLE SAMPLING AND THIN-LAYER CHROMATOGRAPHY KITS

Jeffrey S. Haas and Marjorie Auyong Gonzalez
Lawrence Livermore National Laboratory, 7000 East Avenue, Livermore, CA 94550 USA
Phone: 925-422-6323; E-mail: haas1@llnl.gov; Phone: 925-423-5630; gonzalez3@llnl.gov

Abstract: Sustainable future use of land containing unexploded ordnance requires extensive field assessments, cleanup, and restoration. The ordnance is generally semi-exposed or buried in pits and, because of aging, needs to be handled with caution. Being able to characterize the ordnance in the field to minimize handling, as well as to distinguish it from inert mock material, greatly facilitates assessments and clean-up.

We have developed unique sample preparation methodologies and a portable thin-layer chromatography (TLC) kit technology for rapid field screening and quantitative assessment of stabilizer content in propellants and, energetic materials (explosives) in environmental scenarios. Major advantages of this technology include simultaneous chromatography of multiple samples and standards for high sample throughput, high resolution, very low detection limits, and ease of operation.

The TLC kit technology, sponsored by the Defense Ammunition Center (DAC) of the U.S. Army, is now patented and has been completely transitioned to our commercial partners, Ho`olana Technologies, located in Hilo, Hawaii. Once fully deployed in the field, the new technology will demonstrate a cost-effective and efficient means for determining the percent of effective stabilizer that is remaining on-site and at munitions clean-up sites, as well as munitions storage facilities. The TLC kit technology is also readily applicable for analysis at military or commercial facilities, for a variety of emergency and non-emergency scenarios, and for situations where public concern is high.

RAPID ON-SITE ENVIRONMENTAL SAMPLING AND ANALYSIS OF PROPELLANT STABILIZERS AND THEIR DECOMPOSITION PRODUCTS BY PORTABLE SAMPLING AND THIN-LAYER CHROMATOGRAPHY KITS

Key words: Explosives analysis, field test kit, propellant stabilizer analysis, thin-layer chromatography, defense clean-up, environmental field test kit

1. TECHNICAL APPROACH

1.1 Background:

For onsite analysis, the examination of the vast number of samples necessitates the use of quick, reliable, field portable equipment that can rapidly, quantitatively verify the many chemically different types of ammunition, explosives, and pyrotechnics. The most common suite of analytes to detect is large, consisting of very chemically different compounds and usually occurs at trace levels in complex environmental matrices. This suite encompasses smokeless powders, black powders, and numerous propellant and energetic formulations. Detection should also be sought for common decomposition products of these explosives such as the methylanalines, aminonitrotoluenes, nitrotoluenes, mono- and dinitoroglycerines, and the nitrobenzenes under on-site conditions.

Selection of on-site analytical techniques involves evaluation of many factors including the specific objectives of this work. Numerous instrumental techniques, GC, GC-MS, GC-MS-TEA, HPLC, HPLC-MS-MS, IR, FTIR, Raman, GC-FTIR, NMR, IMS, HPLC-UV-IMS, TOF, IC, CE, etc., have been employed for their laboratory-based determination. Most, however, do not meet on-site analysis criteria, (i.e., are not transportable or truly field portable, are incapable of analyzing the entire suite of analytes, cannot detect multiple analytes compounded with environmental constituents, or have low selectivity and sensitivity). Therefore, there exists no single technique that can detect all the compounds and there are only a few techniques exist that can be fielded. The most favored, portable, hand-held instrumental technique is ion mobility spectrometry (IMS), but limitations in that only a small subset of compounds, the inherent difficulty with numerous false positives (e.g., diesel fumes, etc.), and the length of time it takes to clear the IMS back to background are just two of its many drawbacks.

While conventional TLC analysis is not considered an instrumental technique, it is routinely used in analytical laboratories worldwide for semi-quantitative and qualitative characterization of unknowns. This laboratory-based technique is ideal for rapid screening, is highly sensitive, and is selective for the identification of analytes sought Analytes commonly detected in complex samples are explosives, drugs, plant extracts, pesticides, counterfeit inks, plasticizers, and many other types of organics, organometallics, and propellant stabilizers.

Unlike column chromatography approaches that can only process single samples sequentially, one TLC plate can accommodate and analyze multiple samples and standards. Numerous samples are processed simultaneously in a unique solvent tank, separating out the stabilizer or explosive analyte(s) away from the sample matrix. Semi-quantitative assessments with nanogram detection limits are readily obtained by inspection of the plates. This process allows numerous samples to be analyzed by a single operator per day.

Once the chromatography is complete, only the resolved propellant stabilizer or explosive components appearing as separated spots on the TLC plate can be further enhanced with a unique reagent (See Figure 1). Quantitative analysis and data archival may be performed using an illumination box, camera, and data acquisition equipment as described later. The major advantage of this technology is simultaneous chromatography of multiple samples and standards, extremely low detection limits, potential for quantitation, and its simplicity to operate. Once implemented, the new technologies will be very cost-effective, fast, and efficient for low nanogram detection of propellant stabilizer compounds on-site at military and former military training ranges in a variety of emergency and non-emergency scenarios.

Single Method for All Propellant Formulations

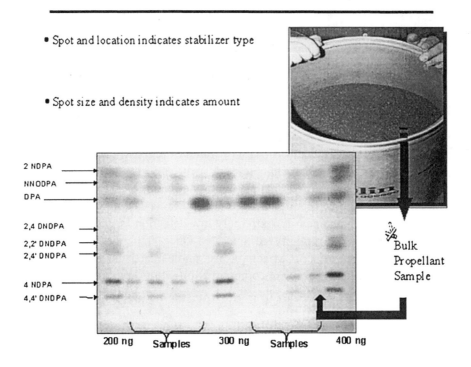

Figure 1.

TLC is also one of the analytical techniques that is commonly used to support evidence in courts of law. The chemistry concept for visualization is not unique since this reaction scheme, converting explosive to pink dyes, is decades old. TLC provide rapid screening capability for the presence of a broad range of explosive residues. TLC also provides a means for obtaining specificity, i.e., identifying numerous types of explosives, their concentrations, and also provides the capability to ratio the amounts of the explosives present. For example, Comp B has a mixture of RDX and TNT in its formulation, and if present in the sample the ratio would be 60:40, respectively. This ratio becomes visually apparent by the density of the spots with TLC technology (See Figure 2).

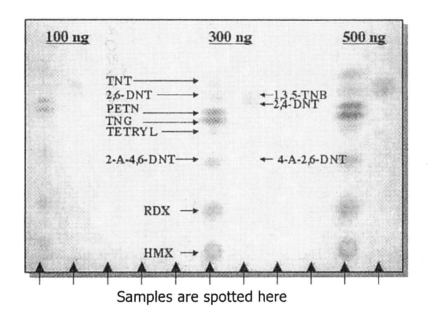

Figure 2.

In Figure 2 above, the three columns of spots are actually three sets of common explosives one may encounter in a very complex sample. For clarity in this picture, only the center column of spots was labeled. Each spot, i.e., each explosive migrates up the TLC plate away from where it was originally spotted, up to a location on the TLC plate that is unique for that particular explosive. HMX is the first spot at the bottom; next one up is RDX, and so on. The density of the spot is related to its concentration. To further show the density, each explosive set was spotted at three different concentrations, the lowest on the left and the highest on the right from left to right at 100, 300, and 500 nanograms.

1.2   Site Evaluation

We have also developed unique sample preparation methodologies and portable TLC kit technologies for rapid field screening and quantitative assessment of explosives in the environment or in propellants (Figure 3). Indicative descriptors are logged throughout the analysis process, ensuring efficiency a higher level of confidence in the reporting of data. Some of the common observations to log are; date, time, conditions, names, and assignment of team members, GPS location, etc. Sample observations are sample sketch, sample type, GPS location of sample, quantity of sample retrieved, sampling method and equipment used. Unusual conditions, difficulties in collection, weather; topography, arctic, jungle, desert, etc. are noted.

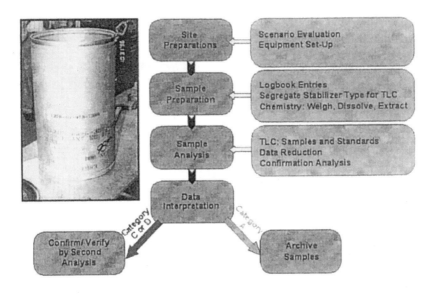

Figure 3.

1.3   Sample Quantity to Collect

The sample quantities to be collected should be sufficient for a replicate analysis. However, alternative plans for sampling can be implemented. For example, a homogenized sample can be divided for on-site analysis, archival, and also to have a portion sent for orthogonal analysis to an off-site lab.

1.4   Chain of custody

Maintaining a chain-of-custody is necessary to ensure proper identification and tracking of each propellant sample from its collection to its analysis. The general procedure is to label samples, record in the sample collection logbook, and then upon transfer to the analyst, the analyst or recipient signs the collection logbook as well as the person releasing the sample. In the haste of an emergency collection scenario, chain-of-custody concerns tend to have low priority. However, sample tracking can be jeopardized and disorganized to the point of becoming a serious issue.

2.   ANALYSIS

Easy Process

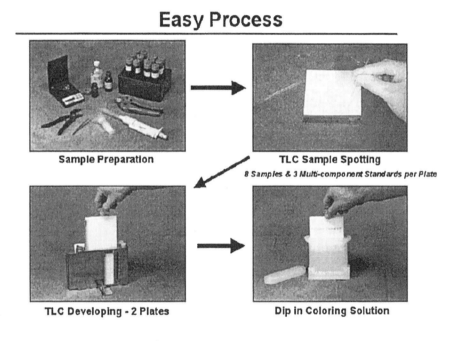

Figure 4.

As shown in Figure 4, small (100-mg) propellant or soil samples are weighed on a portable battery operated balance and placed in 20-ml size vials for dissolution. The solvent system of choice for dissolution is an acetone and environmentally safe HFE 7100 made by 3M. Once the samples are processed, they are ready for TLC analysis.

Pre-packaged, pre-marked TLC plates are then removed from the sealed plastic bag to which 5 uL of each sample and standard are spotted. The concentration of the standards can represent calibration curves over large dynamic ranges or a more critical range close to the detection limits. Each prepackaged TLC plate has eight pre-marked sample locations and three pre-marked locations for standards.

One to two spotted TLC plates are positioned in a TLC rack, and then placed in a unique developing chamber containing a very small amount of solvent. The solvent wicks up the TLC plate, over and past the spotted propellant samples and standards. Chemical components from each spotted sample and standard begin to separate (i.e., chromatography) moving up the plate, and continue to different heights on the TLC plate.

Once the timed chromatographic process is over, the components (energetic materials and associated biodegradation and/or decomposition products) in each sample are then read directly as spots on the TLC plate. The spots are visualized as they fluoresce or cause fluorescent quenching under UV-lamp illumination, or the spots are chemically developed *in situ* using reagents unique to those analytes. As shown in Figure 5, the density and size of the spot are related directly to concentration. However, the visual semi-quantitation and/or quantitation of these separation data can be subjective because of random spot sizes and/or irregular shapes.

New coloring reagents significantly enhance the detection of the stabilizer compounds

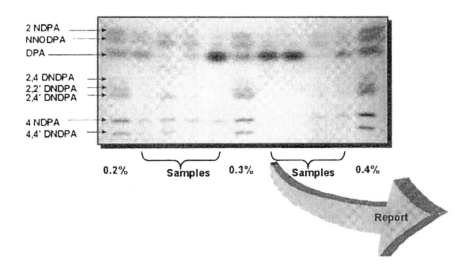

Figure 5.

To increase the objectivity of reading TLC plates, we have designed a unique digital imaging box wherein lies the developed TLC plate. The digital imaging box is complete with UV and room light illumination capabilities and is fitted with a detachable digital camera. The imaged TLC plate is imported into a software program for data reduction.

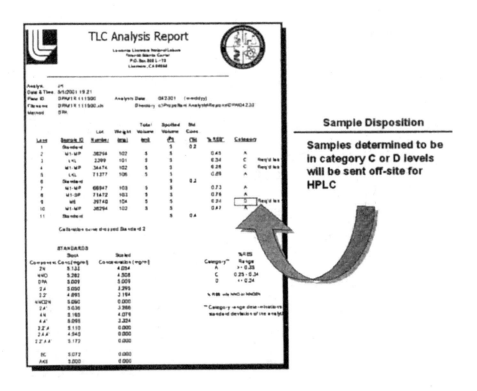

Figure 6.

Extracting the semi-quantitative/quantitative information from a TLC plate requires integration of spot density. The digital camera images (i.e., captures) very low-light spot intensities with three different color channels: red, yellow, and blue. The spots are imaged over time (1/60 of a second) and are conveniently stored on a floppy disk or smart stick and then downloaded directly into the computer. The software integrates the intensity volume for each imaged spot using the best color channel(s) combination and compares it to known standard compounds that were run with the samples on the same plate, thereby analyzing the sample. The propellant's stabilizer content is readily viewed on-screen or printed out in the final report (See Figure 6).

A complete, highly detailed, sample preparation and analysis procedure is provided in the Operations Manual and is available.

3.   SUMMARY

There was a need to develop field-portable, AC power independent, rugged sample collection and TLC kits that are cost-effective, and efficient for analyzing very large numbers of propellant samples per day for stabilizers and explosives on-site at military facilities.

Lawrence Livermore National Laboratory's (LLNL) Forensic Science Center (FSC), is continuing to develop, test, and incorporate state-of-the-art technologies for the portable TLC system. This year, we have further improved the accuracy and precision of the data by developing new technologies, new chemistries, and software analysis routines. The new TLC technology is in kit form and has been transitioned to Ho`olana Technologies, located in Hilo, Hawaii, for the manufacturing of future Army kits. This technology is now also directly applicable, for the rapid sampling and analysis for trace level detection of energetic materials in large numbers of environmental samples.

COMPARATIVE MULTI-ELEMENTAL ANALYSIS OF MINERAL AND ECOLOGICAL COMPONENTS OF THE ECOSYSTEM "KARABASHTOWN" CO-EXISTING WITH A COPPER-SMELTING PLANT. PHYSICOCHEMICAL POINT OF VIEW

N.M. Barysheva[#], N.V. Garmasheva[#]), E. V. Polyakov, V.T. Surikov,V.N. Udachin[&)]
Institute of Solid State Chemistry UB RAS, 91 Pervomajskaya, 620219 Ekaterinburg, Russia, Polyakov@ihim.uran.ru;#)Russian Federal Nuclear Center VNIITF, Snezhynsk, Russia, &)Institute of Mineralogy UB RAS, Miass, Russia

Abstract: Analysis of adequate physicochemical mechanisms of chemical contamination of the territory due to the impact of a copper-smelting plant; comparative statistical analysis of trace elements abundance in the mineral (air, snow, well-water) and biological (potatoes, children's hair) fractions of the Karabash ecosystem in the period of 1997- 2000 are discussed.

Key words: Contamination, physicochemical mechanisms, copper smelter, Karabash ecosystem.

1. INTRODUCTION

Peculiar industrial history and the geographic position of Karabash as a town of rather isolated local industry within the industrially saturated Urals region have made it an object of intensive ecological investigation in the framework of the international project: "The Assessment of Priorities for Middle Urals' Environmental Pollution Prevention". The project was carried our under the aegis of the International Scientific and Technological Center (ISTC). One of the tasks of this project was to find out quantitative indicators of chemical impact of the local industry on the environment.

Within that task, the following sub-tasks were considered in detail for the Karabash ecosystem in the period of 1997-2000:
– internal and inter-laboratory quality control of primary analytical data collected on the territory of Karabash;
– search for and analysis of adequate physicochemical mechanisms of chemical contamination of the territory due to the impact of a copper-smelting plant;
– comparative statistical analysis of trace elements abundance in the mineral (air, snow, well-water) and biological (potatoes, children's hair)

2.   METHODS AND MATERIALS

Sampling procedures were performed in accordance with the grid uniformly covering the territory of the town. Three independent groups of analysts were involved in elemental analysis of snow (74 samples), drinking water (60 samples), soil (21 samples), potatoes (100 samples), and children hair (200 samples) samples [1]. The elements analyzed were $C(HCO_3)$, $S(SO_4)$, Cl, Li, K, Na, Co, Ca, Mg, Si, Fe, Mn, Cu, Zn, Ni, Pb, Cd, Al, As, Cr, Se, Sr, Tl, Bi, V, Hg, Mo, Sn, Sb, and Te.

Acidity (pH) and redox-potential (E_h) of melted snow samples were measured by conventional methods using a "Yokogawa pH 81" ionometer. The main cations and anions in the melted snow fraction were determined by titrimetric analysis. In group 1 (Institute of Mineralogy UB RAS), AAS (Fe, Mn, Cu, Zn, Ni, Co, Pb, Cd) and titrimetric analysis (Ca, Mg, S, Cl, C) were performed to determine elements in melted snow and solid residue upon filtrating 1 liter snow sample through a 0.5 micrometer pore size filter.

Groups 2 (Russian Federal Nuclear Center VNIITF) and 3 (Institute of Solid State Chemistry UB RAS) used ICP-MS for determining all the above listed elements in melted snow and surface water. Soil, vegetable, and hair samples were analyzed by Group 3. As the time interval between collection and analysis of samples was short, no chemical conservation of snow melt was used. AAS was performed using a Perkin-Elmer atomic absorption spectrometer, model 3110. ICP-MS was conducted using Elan 6000 (group 2) and Spectromass 2000 (Group 3) devices.

In all the measurements, conventional calibration techniques for individual elements such as Total Quant, Quantitative Analysis, and RapiQuant modes were used. Merck reference liquid materials for ICP-MS were used for multu-isotope calibration of the mass-spectrometers. In the analysis of soil, national reference solid materials of black soil and lake sediments were used. Laboratory and inter-laboratory quality control procedures were based on conventional statistical methods, regression analysis, and Ftaste for comparisons of variance.

As some necessary meteorological data were unavailable, we employed two different techniques to estimate the element abundance in air. Reverse calculations, in the framework of the American program MEPAS, allowed us to find the concentration fields based on experimental and especially adapted meteorological data. The second technique included direct calculations in the framework of the Russian standardized program Ecologist, which took into account the actual chemical composition of copper-smelting production contaminants. Both techniques had some restrictions, mainly insufficient initial information on the sources of contaminants and limited possibilities of the analytical equipment used.

The following tasks have been solved in the framework of the research:
– full-scale multi-element analysis of snow using modern analytical
  equipment;
– internal and inter-laboratory quality control of primary analytical data
  based on inter-laboratory comparative measurements;
– physicochemical analysis of possible mechanisms of chemical pollution of
  the territory of Karabash due to the impact of copper-smelting plant;
– comparative analysis of trace element concentrations in the mineral (air,
  snow, well-water) and biological (potatoes, children's hair) fractions of the
  Karabash ecosystem

The general aim of the research was to estimate the effect of chemical pollution on the environment and health of the population.

3. QUALITY CONTROL

Snow, especially its water-soluble fraction, is one of the most sensitive and informative indicators of mass-transfer in the chain air – soil – drinking water. Therefore analytical data on snow-melt samples were selected for inter-laboratory quality control. Inter-laboratory verification of analytical results estimated in all the groups have shown that relative standard errors for the concentrations of all the determined elements do not exceed (5-15)% in the concentration range 0.01 – 10000 microg/l, which is consistent with the metrological characteristics of the methods employed. All analytical data collected by different groups of analysts were tested for reliability and generalized using a special statistical procedure. The procedure consisted in successive testing the statistical hypothesis on the homogeneity of the variances characterizing the correlation in sets of analytical data.

In accordance with the physicochemical viewpoint on the origin of pollution of the territory of Karabash, the basic hypothesis, H_b, was as follows: the elemental composition of melt snow has the only "global" source, implying the copper-smelting plant situated in the vicinity of the town. This hypothesis means that the chemical composition of snow and snow melt correspondingly was formed as a result of random scattering of elements from a constant contamination source. Side mechanisms of contamination and fractionation are assumed to be absent. If hypothesis H_b is true, there should be a linear correlation between the element composition of snow samples in any two points on the controlled territory. That supposition is defined as hypothesis H_1. The truth of hypothesis H_1 allows one to verify the discrepancy between analytical data by testing hypothesis H_2 on the homogeneity of variances S_{ij}^2, which characterize the linear correlation of analytical data in any two points (i) and (j) on the town territory in accordance with hypothesis H_1. If hypothesis H_1 is false, the copper-smelting plant is not a single source of pollution. If hypothesis H_2 is false, additional local mass-transfer mechanisms exist on the territory. If hypothesis H_2 is true for the results obtained by each group, inter-laboratory discrepancies can be verified by testing hypothesis H_3: "variances" (S_{ij}^2), which characterize the linear correlation of elements abundance in a given point of the territory and are determined by two arbitrary groups (i) and (j), are "homogeneous". If the latter hypothesis is false, the standard errors in analytical data of the groups differ significantly. Otherwise all inter-laboratory data for each sampling point may be generalized, and average concentrations of all detected elements may be estimated. Since the above procedure differs from conventional quality control methods, it would be expedient to verify its reliability using independent data, which were subjected to conventional reliability tests. For this purpose we used published data on trace elements and precious metals in snow samples collected in immediate vicinity of nickel processing plants, Kola Peninsula [2].

4.   RESULTS AND DISCUSSION

Quality control data were used to verify the basic hypothesis H_b concerning the dominant source of contamination on the territory of

Karabash. Successive testing of derivative hypotheses showed that hypothesis H1 is true for any set of analytical data of all groups involved in the investigation [1]. A linear correlation was established between the concentrations of elements in sets of snow-melt samples when comparing then in the logarithmic scale, equation (1)

$$C_j = a[0] + a[1]C_i, \log C_j = b[0] + b[1]\log C_i , \tag{1}$$

where $C_{i,j}$ denote analytical concentrations (microg/l) of elements in samples $K(A)_{i,j}$ with indices i and j selected at two points of the territory with the same indices; a,b[0] and a,b[1] are dimensional coefficient of the corresponding linear and logarithmic equations. An example of typical correlation is given in Fig. 1.

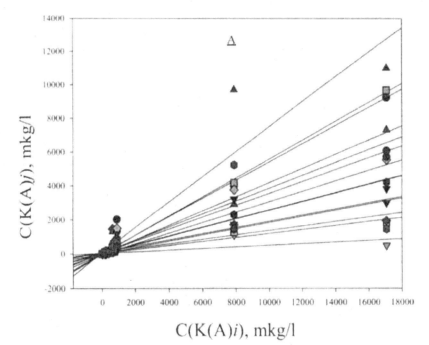

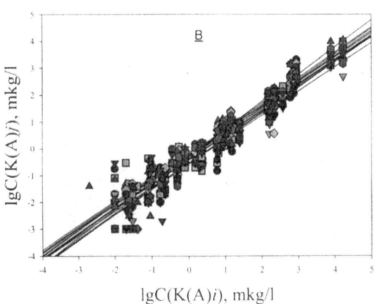

Figure 1. An example of linear correlation between set of elements concentrations (C) in the samples of melted snow from K(A)i to K(A)j, i=27, j=28-41. Estimation is made by ICP-MS.

It should be emphasized that the deviations from the logarithmic correlation obey the normal distribution statistics, allowing one to apply convenient statistical procedures to analytical measurement results. As will be shown below, the logarithmic linear correlation rule was observed for all types of geo-chemical samples, i. e., snow, air, water, and soil. The correlation analysis of the elemental composition of melted-snow fractions showed with confidence level 95% that no significant discrepancy exists between the element composition within the correlation curves (see Table 1) and the corresponding variances are thus homogeneous for any two randomly chosen samples (i. e., points of the territory).

Correlation equation parameters (1)	Samples K(A)i- K(A)j (i=27, j=29)	Samples K(A)i- K(A)j (i=27, j=47)
b[0]	-0.173 ± 0.048	-0.303 ± 0.030
b[1]	0.92 ± 0.03	0.96 ± 0.04
$t_{b[0]}(f)$	-3.6(47)	-4.81(38)
$t_{b[1]}(f)$	33.5(47)	26.7(38)
p-level(b[0])	0.00085	0.00020
p-level(b[1])	0.00000	0.00000
R	0.980	0.974
$S^2_{ij}(f)$	0.100	0.120
$S^2_i(f) / S^2_j(f)$	-	1.20
$F(f, f)_{0.05}$	-	$1.71 > S^2_i(f) / S^2_j(f)$

Index *t* is Student's *t* distribution parameter; *R* is the correlation coefficient, *f* is the degrees of freedom of a sample.

Table 1. Typical statistics of linear correlation of the elements concentration in pairs snow-melt samples.

As can be seen from Table 1, the estimated coefficients b[0] are not equal to zero for different samples, whereas the estimated coefficients b[1] are close to 1 within confidence interval. That means that coefficients b[0] estimated for different points of the territory are generalized relative characteristics of elements abundance at the chosen sampling points. Statistical analysis has confirmed that hypotheses H_1 and H_2 are true with 95% confidence level for the data obtained by any of the analytical groups involved. This conclusion allowed us to verify hypothesis H_3 considering that the estimated average variances of the correlation equation (1) are homogeneous for all snow samples in each analytical group. Hypothesis H_3

was tested using analytical data obtained by all three groups for the same samples.

Both correlation and variance analysis results showed that the hypothesis on the linear correlation between inter-laboratory data and the homogeneity of the corresponding variances is true for all data sets, at the for 95% confidence level. Table 2 presents a typical example of such a comparison. Based on the detected property of homogeneous variances, root-mean-square standard deviation, S, for all melted snow samples was estimated $S = 0.32 \pm 0.06$ for 95% confidence level [3].

Parameters of the correlating equation (1)	Group 1		Group 2		Group 3	
$b[0]$	-0.06 ± 0.26		-0.17 ± 0.05		0.06 ± 0.06	
$b[1]$	1.01 ± 0.10		0.92 ± 0.03		0.92 ± 0.03	
$t_{b[0]}(f)$	$-0.23(10)$		$-3.6(45)$		$-0.96(40)$	
$t_{b[1]}(f)$	$10.4(10)$		$33.5(45)$		$27.4(40)$	
p-level($b[0]$)	0.83		0.00085		0.34	
p-lepel($b[1]$)	0.00000		0.00000		0.00000	
R	0.960		0.980		0.970	
$S_y(f)$	$0.130(10)$		$0.100(45)$		$0.14(40)$	
$S_y(f)$ $S_y(f)$	1.30	1.08	1.3	1.4	1.08	1.4
$F(f, f)_{0.95}$	2.40	2.07	2.40	1.69	2.40	1.68

Table 2. Example of the between-laboratory statisical analysis data for the melted-snow samples K(A)27-K(A)29

This value characterizes the upper level of relative scattering of estimated elements concentration in all the considered snow samples, which is associated both with estimation errors in groups 1-3 and natural variation of the elements abundance in samples. Thus, the results of successive testing of ypotheses H_1 -H_3 allow us to conclude that the basic hypothesis H_b is true and only "global" source of chemical contamination exists on the territory of Karabash.

To make this conclusion more reliable, we applied the above statistical comparison procedure to independent data on the chemical composition of melted snow samples, which were collected in the vicinity of a nickel-processing plant, on the Kola Peninsula [2], and an industrial megalopolis [4]. According to these data, chemical analysis of melted snow was conducted by ICP-MS, ICP-AES and ion chromatography using certified reference materials SLRS-2 of the National Research Council (Canada) and NIST, 1643c (US) [2]. Statistical analysis of these data revealed that hypotheses H_1 and H_2 are true for the chemical composition of any pairs of samples [2, 4]. Analyzing pairs of melted-snow samples randomly chosen in Karabash and the Kola Peninsula [2], we have found the homogeneity of the linear correlation variances in such different objects at 95% probability level. These results allow us to conclude that the analytical data obtained by groups 1-3, as well as the data published in [2, 4], have the same confidence level, and that the observed linear correlation variances reflect a general physicochemical property of chemical elements mass-transfer between the "global" source of pollution and snow.

The abundance of chemical elements in the snow covering the territory of Karabash estimated in three different laboratories is characterized by standard deviations, which correspond to the international specification of quality. The next research step consisted in physicochemical identification of the detected "global" source of chemical pollution on the territory of Karabash. Identification was carried out on the basis of the above information on the chemical composition of melted-snow samples. It was obvious from the spectrum and relative abundance of chemical elements in the snow samples that the most probable source of chemical contamination should be associated with the only copper-smelting plant situated on the territory of the town. To establish the dependence between the technology of blister copper production and the chemical composition of snow and other geochemical objects on the territory of the town, we analyzed in detail the principal schema of the technology. In accordance with [5-7], the blister copper technology includes following principal stages: roasting of copper concentrate, copper-iron sulfide (matte) production, and conversion of liquid matte into the blister-copper melt. The principal flow of chemical elements into the atmosphere (air and snow) can include fine oxides and sulfides (a) and metals (b). The sources emitting (a) include Cu-concentrate, flux, slag, and matte at the roasting and conversion stages. Emission of atoms of metals and their clusters (b) may take place only from the melted copper phase at the final stage of blister copper production.

The following hypotheses was tested in the first approximation: if the vaporization of volatile oxides, sulfides, and metals of all the considered chemical elements at roasting and/or conversion temperature plays a significant role in the contamination of Karabash atmosphere, their calculated equilibrium pressure over the Cu-concentrate, slag, matte or copper melt (or their chemical composition) should strongly correlate with the detected abundance of these elements in snow samples. If such a significant correlation is detected, the corresponding process exerts primary impact on the emission of chemical elements into the atmosphere and may be considered as the predominant source of chemical pollution. Such objects as Cu-concentrate and oxide slag reveal no significant correlation with the composition of melted-snow samples. A weak correlation between the chemical composition of matte dust and snow (average correlation coefficient RMatte=0.445) may probably be attributed to the inclusion and partial dissolution of dust particles in melted snow. The most significant correlation in terms of statistics is that between the abundance of chemical elements in melted-snow samples on the one hand and the calculated equilibrium pressure of individual metals over their melts at the temperature of blister copper formation on the other hand, Fig. 2. The strong correlation between the measured chemical composition of snow and the calculated pressure of individual metals in the melted copper, allows us to conclude that the blister copper formation process is most likely to be the critical mechanism of pollution of the atmosphere and the territory of Karabash.

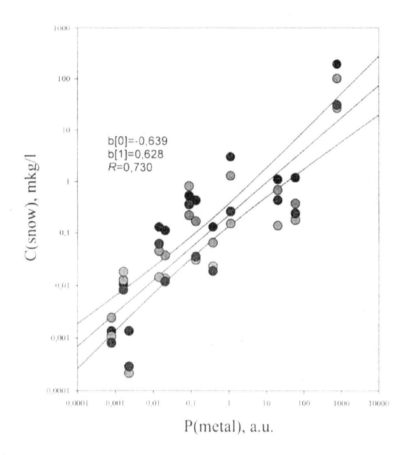

Figure 3. An example of correlation between the abundance of low volatile metals in a random set of Karabash snow samples (C) and the calculated volatility of the same pure metals at the temperature of blister copper formation (12600 C) . Vertical points correspond to the selected samples K(A)j; the data for Si, Cu, and Fe are not considered.

The above inference concerns all chemical pollutants except Fe, Cu, Si, and Ni. The abundance of these elements in melted-snow samples is beyond the limits predicted by the above model of vaporization and consequently can be attributed to non-molecular forms of mass-transfer of these elements. This discrepancy can be explained by some additional sources of contaminants within the considered technology. A comprehensive comparative analysis shows that the most probable form of transferring such elements as Fe, Cu, Si, Ni into the atmosphere and snow are the matte and dust, where they are major chemical elements. A rather strong correlation between the abundance of those chemical elements in the matte dust and snow supports this supposition. Thus, generalizing the snow contamination mechanism we may conclude that the blister copper production technology is a dominating factor of chemical contamination. To be more exact, the transfer of pollutants occurs on the final stages of conversion and blister copper formation. As the principal source of chemical contamination has been identified, it was important to trace its effect on other objects of the Karabash geo- and ecosystem. The samples of soil, drinking water, air, potatoes, and children hair were selected as most informative objects of the investigation. We have determined the origins of soil and water contamination on the territory of Karabash caused by man. This allowed us to find out the key role of contaminants accumulation in the soil and well water, subsequent inclusion of contaminants into food chains and assimilation by the population.

Using the comparative multi-element correlation analysis we have shown that the soil is now almost completely saturated with heavy metals and cannot serve as a natural geochemical barrier. Therefore the contamination of the environment inevitably results in increased elements abundance in well water. This increases the involvement of heavy metals in food chains. Analysis of chemical elements distribution in the system "soil – plants" indicates that plants differ significantly in the accumulation of elements. It is shown that potato is the most "harmless" crop the population uses. In contrast, onion is found to be the most harmful crop due to intensive accumulation of heavy metals from the soil. Analysis of hair samples revealed a strong correlation between the abundance of such elements as As, Hg, Pb, Cd in well water on the territory of the town and in the hair of children in the age of 4-7 years. It was proved statistically that this correlation reflects the general influence of the main contamination source on the ecosystem of Karabash. In contrast, this factor is statistically negligible for the territories with centralized water supply. This allowed us to formulate practical recommendations for immediate protection of the population. Significant changes in the general watersupply system of Karabash, as well as modifications of the gas-purifying system in the of copper-smelting technology were proposed as countermeasures. This work is fulfilled under support of the ISTC projects Nos. 500 and 1872.

REFERENCES

1. Polyakov, E.V., Surikov, V.T., Bamburov, V.T., Shveikin, G.P., Barysheva ,N.M., Batalova, I.A., Eremkina ,T.V., Udachin, V.N. IX International Symposium " Ural Atomic, Ural Industrial", Ekaterinburg, UB RAS. 2001.123.
2. Gregurek, D., Reimannb, C., Stump, E.F. Trace elements and precious metals in snow samples from the immediate vicinity of nickel processing plants, Kola Peninsula, northwest Russia Environmental Pollution. 102. 1998. 221-232.
3. Standard practice for use of statistics in the evaluation of spectrometric data. ASTM, E876-89.
4. Hoffmann, P., Karandashev, V.K., Sinner, T., Ortner, H.M. Chemical analysis of rain and snow samples from Chernogolovka/Russia by IC, TXRF and ICP-MS Fresenius J. Anal. Chem. 357. 1997. 1142–1148.
5. Kazakov, N.F., Osokin, A.M., Shishkova, A.P. Processing of metals and construction materials. Moscow, Metallurgy, 1975. (in Russian). 6. Melting in liquid bath. / A.V. Vanukov, Moscow, Metallurgy, 1988. (in Russian).
7. Britannica CD 2000 Deluxe Edition./CD-ROM-based Encyclopedia Britannica. 2000.
8. Handbook of Chemist. V1. Leningrad. Khimiya. 1971. (in Russian).

AUTOMATED ANALYSIS OF STABLE ISOTOPES OF H, C, N, O AND S BY ISOTOPE RATIO MASS SPECTROMETRY

Janusz A. Tomaszek
Dept. of Environmental & Chemistry Engineering, Rzeszów University of Technology, 2 W Pola Street, 35-959 Rzeszów, Poland, E-mail: tomaszek@prz.rzeszow.pl

Abstract: This paper reviews the principles and history of gas isotope ratio monitoring focusing on the instrumentation used in isotope ratio monitoring. We review the analytical systems currently available both for traditional dual-inlet isotope ratio mass spectrometry (DI-IRMS) and the newer continuous-flow isotope ratio mass spectrometry (CF-IRMS).

Selected applications of isotope ratio monitoring technique in mass spectrometry to geochemistry, biology, medicine and other areas of science are described. It was shown that gas isotope ratio mass spectrometry has greatly increased our understanding of the biogeochemical cycles of C, N, S and H_2O in natural and agricultural ecosystems.

Key words: Mass spectrometry, stable isotopes

1.   INTRODUCTION

Mass spectrometry is one of the oldest instrumental analytical methods. Positive rays were discovered by Goldstein in 1886 (after Barrie & Prosser, 2000). The first mass spectrometer for routine measurements of stable isotope abundances was reported in 1940 and improved upon over the following ten years Nier, 1940, Nier, 1947, Murphey, 1947, McKinney et al, 1950, after Prosser, 1993. It is remarkable that the vast majority of active gas spectrometers in use today are little changed from those described around 50 years ago. For most people, 'mass spectrometry' now means organic molecular structure determination. However, within the last 15 years, gas isotope ratio mass spectrometry has greatly increased our understanding of the biogeochemical cycles of C, N, S, and H_2O in natural and agricultural ecosystems. Automated analysis of stable isotopes of the elements: H, C, N, O, and S has become highly developed, and there is now a wide array of equipment to meet an ever-growing range of applications.

The purpose of this paper was to briefly describe fundamentals of isotope ratio mass spectrometry (IRMS), review the analytical systems currently available both for traditional dual-inlet (DI-IRMS) and the newer continuous-flow (CF-IRMS) and describe the specialized instruments that are in general use for isotopic measurements.

We gave also a mini-review through different applications of the IRMS technique.

2.   STABLE ISOTOPES

The elements whose isotopes are routinely measured with gas inlet mass spectrometers are carbon (^{12}C and ^{13}C, but not ^{14}C), oxygen (^{16}O, ^{17}O, ^{18}O), hydrogen (^{1}H, ^{2}H, but not ^{3}H), nitrogen (^{14}N and ^{15}N) and sulphur (^{32}S, ^{33}S, ^{34}). Stable isotopes of H, C, N, O, and S occur naturally throughout atmosphere, hydrosphere, lithosphere, and biosphere. They are atoms of the same elements with a different mass. Each element has a dominant 'light' isotope with the nominal atomic weight (^{12}C, ^{16}O, ^{14}N, ^{32}S, and ^{1}H) and one or two 'heavy' isotopes (^{13}C, ^{17}O, ^{18}O, ^{15}N, ^{33}S, ^{34}S, and, ^{2}H) with a natural abundance of a few percent or less Table 1).

Table 1. Natural abundance of stable isotopes (atom %) (Oebelmann et al., 2000)

^{1}H	99.985	^{2}H	0.01557
^{12}C	98.892	^{13}C	1.11140
^{14}N	99.6337	^{15}N	0.36630
^{16}O	99.759	^{17}O	0.03740
		^{18}O	0.20390
^{32}S	95.018	^{34}S	4.21500

It was once thought that the abundances of these isotopes were constant throughout nature. However, as measurement precision improved, small differences were observed in chemically identical materials of different origin, or produced by different processes. Although isotopes of the same element take part in the same chemical reactions, they do so at different rates. Chemical reactions and physical processes such as evaporation and condensation discriminate against heavy isotopes. This 'fractionation' results in products which are isotopically lighter (contain less of the heavy isotope) than their starting materials. This is evident in the hydrologic cycle where snow falling at the poles is depleted in ^{2}H and ^{18}O relative to rainfall at the equator. An understanding of these *natural abundance variations* led to practical applications in geochemistry, oil exploration, and hydrology during the 1950s and 1960s. Later, it was discovered that biological processes such as photosynthesis and N-fixation also fractionated stable isotopes. These natural abundance variations enable a wide array of applications in soil science (Figure 1-4), (Oebelman et al., 2000).

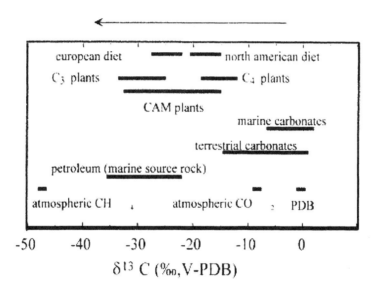

Figure 1. Carbon isotope variation in nature

An alternative approach uses *tracers*, isotopically enriched materials, which are added to the system studied. Labeled compounds are available with the heavy isotope enriched to a level of 99 atom%. These are more difficult to produce than radioisotopes, making them relatively expensive. Materials labeled with stable isotopes are non-radioactive and pose no hazard to human health or the environment (Barrie & Prosser, 2000).

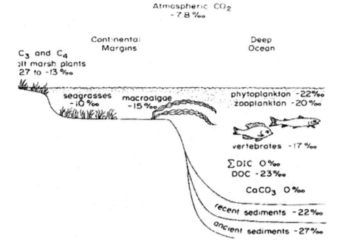

Figure 2. Carbon isotope variation in nature

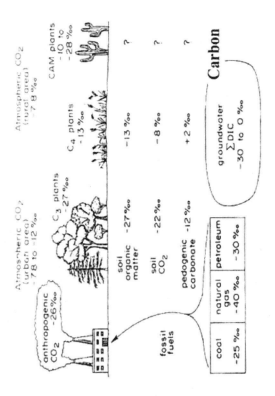

Figure 3. Stable carbon isotope ratios of major components of terrestrial ecosystems

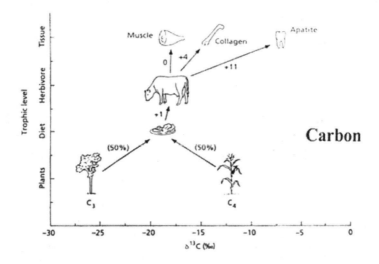

Figure 4. The fractionation of carbon between diet and different tissues in mammalian herbivores

3.   MEASUREMENT NOTATION AND STANDARDS

Isotopic measurements are usually reported in one of three ways:

1. *Abundance* in units of atom% is used for tracer measurements. This is an absolute measurement of the number of atoms of the isotope in 100 atoms of the element. For example:

^{13}C abundance,
$$atom\%\,^{13}C = \frac{100\left[^{13}C\right]}{\left[^{12}C + ^{13}C\right]}.$$

2. *Atom% excess*, written as APE, measures the abundance above a specified background level. For ^{15}N, the value of 0.3663 atom % for N in air is commonly taken as the natural background.

3. Natural abundance data are nearly always reported as *delta values*, δ in units of per mil ("mil" = 1000), written ‰. This is a relative measurement made against a laboratory's own reference material, a "working standard", calibrated against an international standard. Delta values are calculated from measured isotope ratio as:

$$\delta = \frac{1000\left(R_{sample} - R_{s\,\tan dard}\right)}{R_{s\,\tan dard}}$$

were R_{sample} is the ratio of the heave to the light isotope measured for the sample and $R_{standard}$ is the equivalent ratio for the standard.

4.   MASS SPECTROMETERS

A mass spectrometer ionizes gaseous molecules and separates the ions into a spectrum according to their mass-to-charge ratio, m/z, (m-the mass of the ion, z – electrical charge) using electric and magnetic fields. The relative abundances of molecules of different m/z are then found by measuring the currents generated by these separated ion beams. All instruments have three basic parts: an ion source, a mass analyzer, and an ion collector assembly (Fig 5).

Precise measurement of isotopic ratios requires a special type of mass spectrometer. There are important differences between the IRMS and a regular organic mass spectrometer.

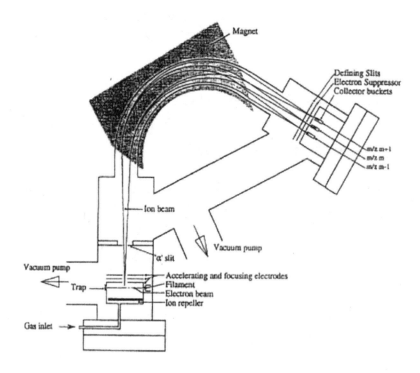

Figure 5. Schematic diagram of an IRMS

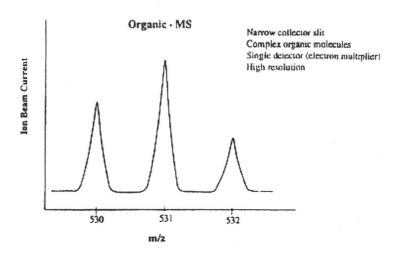

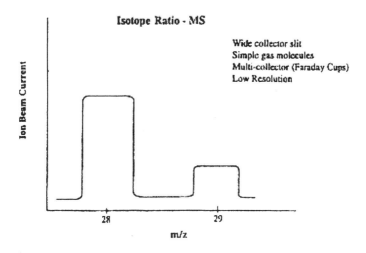

Figure 6. Comparison of mass spectra an features of an IRMS an MS

For both instruments, the sample is ionized by electrons in the source, ion are accelerated down a flight tube between poles of a magnet, and they are deflected in proportion to their mass-to-charge ratio. However, in the organic mass spectrometer, the ion beam is swept over a slit by varying either the accelerating voltage or strength of the magnet spectrum is recorded. In the isotope ratio mass spectrometer, conditions are held constant throughout the analysis and the relative intensity of two or more ion beams are measured simultaneously. IRMS typically cover the m/z range 2-100. IRMS is designed to produce strong (10-11-10-8 A) steady beams in low-noise Faraday cups to give high precision. Samples are simple gases H_2, CO_2, N_2 and SO_2, requiring only low resolving power ($m/\Delta m$ of <100). Resolving power is defined as $m/\Delta m$, where an ion beam at mass m is separated from another of equal intensity at mass $m+\Delta m$ with an overlap between the two peaks of 10% relative to the peak height. Organic mass spectrometers, on the other hand, often need much greater resolution to separate closely spaced peaks at high ass and must accept what little signal is available and the non-linear response of electron multiplier detectors.

So how does the IRMS get its stability ? Collector slits are several times the width of the ion beams. This gives a "flat-topped " peak shape (Fig 6) which makes the ion current intensive to drift. The main source of drift is temperature variation which both affects the electronic components used for mass selection and caused expansion and contraction of mechanical parts. Simultaneous measurement of ion beams using a double or triple collector is more precise than sequential measurement by mass scanning with a single detector. Finally, frequent comparison of sample gas under identical conditions also contributes to stability. Ion beam stability is more important than resolution for isotopic measurements.

Ions move in circular path in a magnetic field as described (Garaj et al., 1981, Namiestnik, 1992, Barrie & Posser, 2000):

$$\frac{m}{z} = \frac{(Br)^2}{2V},$$

where:

          r – radius of the circle,
          B – strength of magnetic field,
          V – accelerated voltage.

From this we see that m/z can be selected by varying either V or B. We will use N isotopes as an example. Referring to Fig. 5, the accelerating voltage between source and analyzer and the strength of the magnetic field have been chosen so that the major $^{14}N_2$ ion with a m/z of 28 follows the path of tightest radius, labeled „$m - 1$";the heavier $^{15}N^{14}N$ ion at m/z -29 goes down the center of the analyzer, labeled „m", whereas the ion beam for $^{15}N_2$ at m/z -30 is deflected least and enters the collector labeled „$m + 1$".

Each ion beam passed through a resolving slit and gives up its charge to the appropriate Faraday collector. This generates an ion current proportional to the rate at which ions are arriving at the collector, and hence to the partial pressure of that isotopic species in the gas sample.

A high-vacuum system keeps the analyzer pressure low enough to reduce collision between ions and background gas to an acceptable level. In the dual mode, analyzer is operated at a pressure below 5×10^{-7} mbar, whereas during the continuous-flow mode, this may be as high as 10^{-5} mbar because of the presence of the carrier gas. At the high pressures, the mean free paths of gas molecules are typically 10-100 times the ion path, through the analyzer. An ion is more likely to collide with a gas molecule, be knocked off its path, and end up in the wrong collector. This appears in the mass spectrum as a tail from the major peak running into an adjacent minor peak. Fig. 7 exaggerates this for clarity. The ratio of this tail to the height of the major beam is called the *abundance sensitivity, ψ*. A systematic overestimate in the measured abundance of the minor isotope is produced and is not corrected procedure of sample-reference comparison. The result is an inaccuracy in the final δ value; the true δ scale is compressed by a factor proportional to the abundance sensitivity:

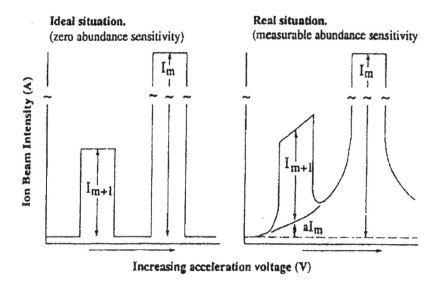

Figure 7. Effect of abundance sensitivity on isotope ratio determination

$$\delta_{measured} = \frac{\delta_{true} R_{reference}}{R_{reference} + \psi},$$

$R_{reference}$ - ratio of heavy isotope to light isotope for the reference

$$\delta_{true} = k\,\delta_{measured}$$

$$k = 1 + \frac{\psi}{R_{reference}}\ , \text{(Prosser, 1993)}$$

Typical abundance sensitivities for modern, differentially pumped instruments measuring CO_2 would be <1 ppm (99.99% accurate for DI-IRMS and ~ 30 ppm (~99.7% accurate) for CF-IRMS. For instruments with only one high-vacuum pump on the analyzer, the figure is approximately an order of magnitude higher; only 97% accurate for CF-IRMS. For N_2, ψ is usually insignificant because of the greater relative difference and greater spatial separation among m/z: 28, 29, and 39 compared to m/z: 44, 45 and 46 for CO_2.

The ability to separate ions spatially is called the *dispersion* of a mass spectrometer. Dispersion is simply the distance between the centers of two ion beams that differ in mass by Δm at the collection plate. A simple sector instrument, where the ion beam enters and exist the magnetic field normal to the pole faces and the object and image distances are the same, is known as a *symmetrical geometry* analyser. Examples are shown in Fig. 8. In this case, the dispersion, D is given by:

$$D = r(\frac{\Delta m}{m})$$

Many older isotope ratio mass spectrometers employed symmetrical geometry. Most new instruments employ asymmetric design known as extended geometry, providing greater dispersion for the same radius (~2x that of a symmetrical geometry).

a.) Symmetric ('Normal') Geometry.

b.) Asymmetric ('Extended') Geometry

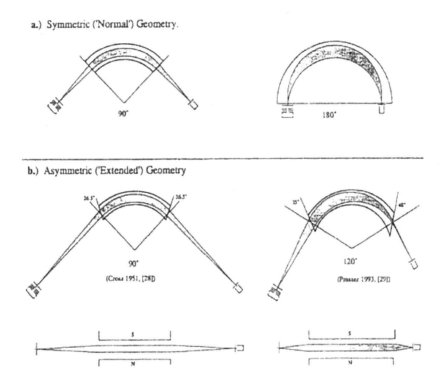

Figure 8. Types of magnetic sector analyzer

5. ANALYTICAL SYSTEMS

5.1 DI-IRMS

For applications requiring very high precision (better than 0.1 ‰) to detect small isotopic differences at the natural abundance level, a dual or triple collector IRMS with a dual-inlet system remains the instrument of choice. A dual inlet (Fig. 9) alternately admits sample and reference gases into the mass spectrometer, perhaps six times each over 10 min, to give the effect of simultaneous measurement. The basic sdesign is similar to the original (McKinney et al., 1950). However, improvements in electronics, vacuum technology, and computer control have meant that inlet systems and data processing are now automated and electronics are more stable.

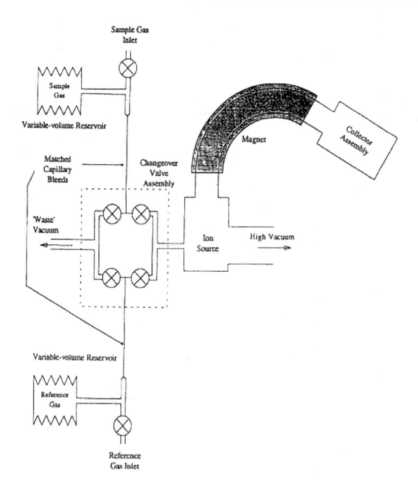

Figure 9. Principles of dual-inlet IRMS

The basic principles of DI-IRMS is a comparison between sample and reference gases in much the same way as is spectrophotometry, where a comparison is made between sample and reference cells. Principles of continuous-flow IRMS are shown in Fig. 9. An automated dual-inlet equalizes sample and reference gas pressures, and hence their major ion beam currents, by adjusting the volumes of two storage reservoirs. The early McKinnney designs used mercury moving in glass tubes as adjustable reservoirs, but modern instruments use flexible stainless-steel bellows. The outlet from each half of the inlet is connected to a changeover valve assembly through "balanced" capillaries.

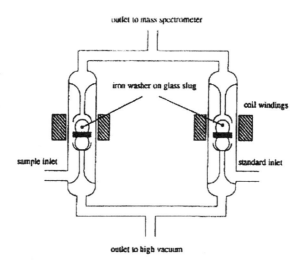

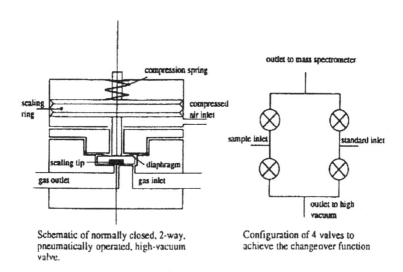

Schematic of normally closed, 2-way, pneumatically operated, high-vacuum valve.

Configuration of 4 valves to achieve the changeover function

Figure 10. The changeover values used by dual-inlet IRMS

This ensures that sample and reference gases both flow from their respective reservoirs at the same rate. As the changeover valve switches, sample and reference gases alternately flow either into the ion source or into a "waste" vacuum line. The original McKinney changeover and modern one are shown in figure 10.

5.2   CF-IRMS

The principles of CF-IRMS is that sample chemistry and gas purification take place in a He carrier to produce pulses of sample-derived N_2, CO_2, N_2O, or SO_2 gases which flow directly into the ion source (Fig. 11).

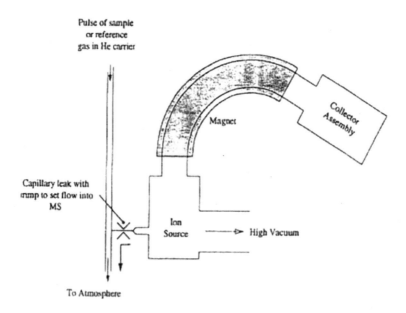

Figure 11. Principles of continuous-flow IRMS

CF-IRMS provides reliable data on micromoles or even nanomoles of sample without the need for cryogenic concentration because more of the sample enters the ion source than in DI-IRMS. CF-IRMS instruments accept solid, liquid, or gaseous samples such as leaves, soil, algae, or soil gas, and process 100-125 samples per day.   Automated sample preparation and analysis takes 3-10 min per sample. The performance of CF-IRMS systems is largely determined by the sample preparation technology. A variety of inlet and preparation systems is available, including GC combustion (GC/C), elemental analyzer, trace gas pre-concentrator and other. The novel

IRMS system configuration is shown in Figure 12. Measurement of H/D ratio described by Hilkert et al., 1999 is one of the novel applications.

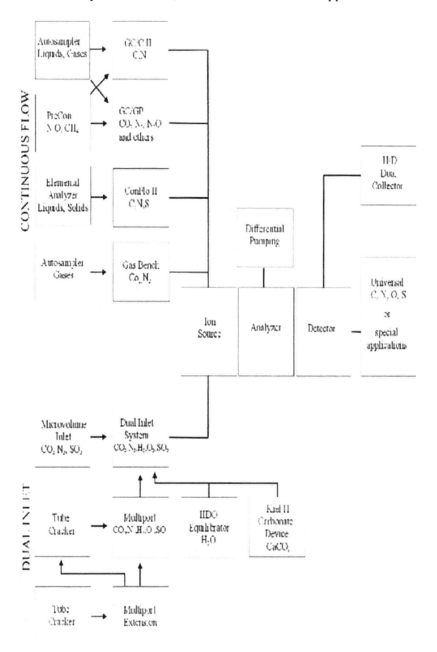

Figure 12. The novel IRMS system configuration

6.   CONCLUSION

There are now a number of options open to scientists who need to analyse stable isotopes as part of their research. The choice is dependent on many factors, such as the type of sample, the precision of measurements required, the amount of sample available, and the number of samples to be analysed. Dual inlet IRMS is still the technique of choice those scientists who need to measure a few samples to very great precision. More often, scientists want to determine a trend in conditions over time, or a flux integrated over an area. In these cases, CF-IRMS will be preferred. Continuous-flow IRMS is also preferred when a limited amount of sample material is available, as it is able to analyze far smaller samples (between 10 and several hundred times smaller) than DI-IRMS.

Isotope techniques provide some very powerful tools for scientists. However, IRMS instruments are relatively expensive and overcomplicated for laboratories with limited applications.

REFERENCES

1. Barrie, A., Prosser, S.J. Automated analysis of light-element stable isotopes by isotope ratio mass spectrometry. Unpublished Materials. Europa Scientific Ltd. Crewe, Cheshire, England, 2000.
2. Brand, W.A. High precision isotope ratio monitoring in mass spectrometry. J. of Mass Spectrometery. 31, 1996, 225-235.
3. Garaj, J. Fizyczne i fizyko – chemiczne metody analizy. W N T, 1981, Warsaw.
4. Hilkert, A.W., Douthitt, C.B., Schluter H.J., Brand W.A. Isotope ratio monitoring gas chromatography/mass spectrometry of D/H by high temperature conversion isotope ratio mass spectrometry. Rapid communications in mass spectrometry. 13, 1999, 1226 – 1230.
5. McKinney, C.R., McCrea, J.M., Epstein, S., Allen, H.A., Urey, H.C. Improvements in mass spectrometers for the measurement of small differences in isotope abundance ratios. Rev. Sci. Instrum. 21,1950, 724.
6. Murphey, B.F. Phys. Rev. 72, 1947, 834.
7. Namiestnik, J. Metody instrumentalne w kontroli środowiska. Gdańsk University of Technology Publisher, 1992, Gdańsk, Poland.
8. Oebelmann, J., Juchelka, D., Hilkert, A., Avak, H., Douthitt Ch. Authenticity control by multiple element isotope ratio determination. Unpublished Materials. Thermo Finnigan, 2001.
9. Nier A.O. Rev. Sci. Instrum. 11, 1940, 212.
10. Nier A.O. Rev. Sci. Instrum. 18, 1947, 398.
11. Prosser, S.J. A novel magnetic sector mass spectrometer for isotope ratio determination of light gases. Int. J. of Mass Spectrometry and Ion Processes. 125, 1993, 241-266.
12. Zimnoch, M. Wachniew, P. (2000). V Isotope Wokshop. July, 1-6, 2000, Krakow, Poland.

6

ENVIRONMENTAL SECURITY

RELATIONSHIP OF ENVIRONMENT AND SECURITY

Katarina Mahutova and John J. Barich
U.S. Environmental Protection Agency Region 10, 1200 6th Ave (OEA-095), Seattle, WA 98101, E-mails: mahutova.katarina@epa.gov, barich.john@epa.gov

Abstract: New definition of national security has emerged. Non-military threats, such as environmental mismanagement, natural resource depletion, overpopulation and the environmental consequences of the Cold War, economic decline, social and political instability, ethnic rivalries and territorial disputes, international terrorism and drugs have been included into the definition of national security. The relationship between environment and security has been under consideration almost two decades. Although nations continue to be central actors in international politics, they increasingly participate and are engaged in cooperation with international and regional organizations to respond to non-traditional security concerns including the environment.

The effort of this presentation is to assess the links between environment and security based on conceptual model. This model will show the different factors, which have the influence on the relationship between the environmental change and security as the triggers, catalysts or interactors.

These factors help to identify the general types of environmental conflicts. According to this typology of environmental conflicts, there is a large potential for local, regional and international collaboration in the various policy arenas in order to establish the coherent mechanisms to environmental conflict.

1.   INTRODUCTION

It has been recognized that security is not entirely a function of military power or geopolitical strength. The security organizations and institutions in other policy areas are increasingly concerned with non-traditional threats to security. A new definition of national security has emerged. It has been broadened to include the threats in the form of environmental mismanagement, natural resource depletion, overpopulation, and the economic consequences of the Cold War, as well as economic decline, social and political instability, ethnic rivalries and territorial disputes, international terrorism, and drugs. The broader security concept recognizes that security and stability have political, economic, social and environmental elements.

Although nations continue to be a central actors in international politics, they increasingly participate and are engaged in cooperation with international and regional organizations (e.g., NATO, UN, EU, etc.) to respond to non-traditional security concerns including the environment.

2.   METHODOLOGY

This research presents a conceptual model describing the links between environment and security. This model uses the different factors (i.e., triggers, catalysts, or indicators) that influence the relationship between environmental changes and security.

The degree to which environmental stress actually contributes to the incidence and escalation of conflict depends on the relationship between the consequences of environmental stress and on number of socio-economic, political and other contextual factors. If these contextual factors are unfavorable, the incidence of conflict due to the consequences of environmental stress is likely. If the contextual factors are favorable, the probability of a peaceful solution is improved.

3.   RELATIONSHIP BETWEEN ENVIRONMENT
AND SECURITY

The relationship between environment and security has been under consideration almost two decades in both the scientific and policy communities.

Environmental and security issues are multifaceted and complex, in cultural, political and scientific sense. In a fundamental way, the environment must be viewed as a strategic factor to be weighed in with many other variables affecting a global or homeland security.

The global integration of human activities, and the realization of the importance of environmental factors in achieving and sustaining a desirable quality of life, have added the complexity to this picture. Such developments require of us more complex approaches to the maintenance of our own security.

Environmental stress (e.g. natural resource depletion, etc.) is one of the elements having an influence on security. The relationship is so complex and interrelated.

In the NATO CCMS Pilot Study on Environment and Security in an International Context, (1999) the relationship between environment and security has been characterized by:

– Multi-causality: environmental stress contributing to conflict almost always interacts with other political and economic factors and evolves through various multi stages before it results in conflict; This is depicted in Figure 1.

Reciprocity and feedback loops: the relationship between environmental stress and conflict is recursive; because just as environmental stress can lead to conflict, conflict can lead to more environmental stress. See Figure 1.

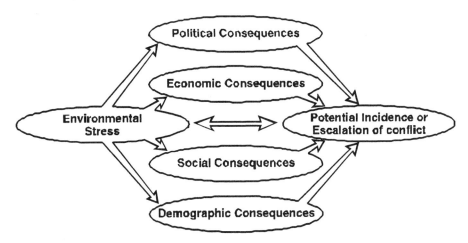

Figure 1. Links between Environmental Stress and Conflict

Consequences of environmental stress: poverty, food insecurity, poor health conditions, displacement (migration or refugee movements) and disruption of the social and political institutions are regarded as the most important consequences from environmental stress, which then contribute to conflict under a certain set of unfavorable factors.

Environmental stress can also behave differently, depending of factors. It can be a structural source of conflict, catalyst or trigger for conflict. It depends also of:

– the gravity of the stress,
– the probability that the stress has actually been realized,
– the duration of the stress and

  − the timing of onset of the stress.

  Similar types of environmental stress may have different effects on security. Therefore the socioeconomic and political context, in which the environmental stress occurs, has to be taken into consideration while assessing the potential of the conflict of different environmental stress. The factors influencing the relationship between environment and security are so called contextual factors. They can also have either a facilitating or inhibiting effect on this relationship.

  The figure shows that contextual factors influence whether environmental change causes, social, economic and political consequences which in turn impact on security.

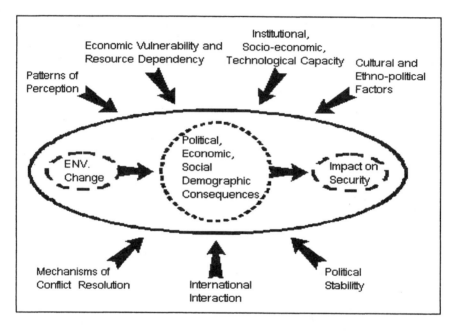

Figure 2. Contextual model

  − Patterns of perception
    Whether or not environmental stress contributes to the potential incidence or escalation of conflict depends heavily upon the perception of the actors. If environmental stress has an impact on physical or economic well-being, actors are more willing to escalate the potential for conflict.
  − Economic vulnerability and resource dependency
    Developing countries are at higher risk because they are more directly dependent on their natural resources. Dependence differs enormously among countries, but also among regions and social groups.
  − Institutional, socio-economic, technological capacity

The creation, distribution and application of knowledge can be seen as a precondition for the environmental stress occurrence minimization and to preventing potential conflicts.

– Cultural and ethno-political factors

The existence of ethnic, cultural or religious differences within a state do not in themselves cause a conflict, but they contribute to the incidence or escalation of conflict, if they develop into a political problem

– Political Stability

The potential incidence of conflict is high in the presence of political instability

– International Interactions

Regional and international interactions are likely to reduce the incidence of conflict. They help cooperatively resolve the negative consequences of environmental stress. To solve environmental problems can help advance the goal of political stability, economic development, and peace.

Lot of agencies and institutions have recently developed partners in environment and security on regional, national and global level among departments, national laboratories, governments, associations, research institutions

– Mechanisms of Conflict Resolution

The presence of effective and legitimate legal, political and social mechanisms of conflict resolutions enhances the possibility of resolving conflict.

Environmental factors and security are tightly linked. Each affects the other. Because of multiple factors, linkages, and causalities, the relationships are complex and often difficult to understand. The use of a conceptual model, such as the one suggested in this paper, can help in managing the resolution of security issues caused by environmental stress.

INTEGRATED RISK ASSESSMENT AND SECURITY

Katarina Mahutova and John J. Barich
USEPA Region 10, 1200 6th Ave (OEA-095), Seattle, WA 98101, E-mails: mahutova.katarina@epa.gov, barich.john@epa.gov

Abstract: It is difficult to assess the risk to security caused by environmental change because the relationship between environmental stress and security is indirect. Environmental stress may cause a series of consequences such as political, economic, social and demographic, and these consequences impact on the potential incidence or escalation of conflict. As we know, the environmental, economic and social issues are interdependent and cannot be pursued separately. The purpose of this effort is to introduce the integrated risk assessment as a practical tool for identifying the potential conflict and consequently the threat to security.

1.   INTRODUCTION

The integrated risk assessment takes into consideration the broad range of factors such as the political, economic, social, demographic and environmental factors. The integrated risk assessment helps to control or manage the complexity of the relationship between the consequences of the stress and the factors caused the stress in order to determine which factors have the most potential impact on the incidence or escalation of the conflict. In the integrated risk assessment the using of the syndrome approach (a set of pre-established patterns of interactions) can help to control the complexity. The syndrome-based concept starts from the assumption that environmental stress is a part of human - natural interactions, which is very dynamic. The Syndrome Approach identifies the different types of interactions, which occur in various environmental or geopolitical regions of the world.

Fourteen experimental hypotheses divided into the three subgroups of 'resource use', 'development' and 'sink' are presented as Tables 1, 2 and 3.

Using the integrated risk assessment allows the identification of when, where and how a syndrome (represented as a stress) might lead to conflict.

2. INTEGRATED RISK ASSESSMENT, A TOOL TO ASSESS THE IMPACT OF ENVIRONMENTAL STRESS ON SECURITY

The relationship between environment and security is indirect and multi-casual. Therefore it is difficult to assess the risk to security caused by environmental change. Taking into consideration the principle of sustainable development where the environmental, economic and social issues are interdependent and cannot be pursued separately, the integrated risk assessment is particularly useful.

The assessment of risk of the potential incidence or escalation of conflict is called integrated because of the broad scale of factors that are considered (political, economic, social, demographic and environmental. The integrated risk assessment manages the complexity in the relationship between the consequences of stress and inhibiting or facilitating factors in order to determine which factor has the most potential impact on the incidence or escalation of conflict.

Using a set of pre-established patterns of complex interactions, or syndromes, can help to simplify the complexity of the relationship of environment and security.

The Syndrome Approach developed by the German Government's Advisory Council on Global Change (WBGU, 1997) and the Potsdam Institute for Climate Impact Research (PIK) provided a set of experimental hypotheses as templates for pattern matching which helps to control the complexity in the integrated risk assessment. The research on the climate, atmosphere, hydrosphere, soil, biodiversity, population, migration, urbanization, economics, societal organisation, psychosocial sphere and technology was taken into consideration to develop the set of these patterns.

For the purpose of this paper the set of experimental hypotheses from WBGU (1997) was adopted and modified (see Table No.1, 2, 3 and 4).

Tables 1 through 3 provide the description of fourteen patterns of environmental stress along with the symptoms. The core problems caused by individual   syndromes   are   visually   entered   into   Table   No.   4.

Table 1. Patterns of environmental stress: Resource-Use Syndrome

Resource Use Syndromes	Description	Symptoms
Overcultivation Syndrome	Overcultivation of marginal land	Destabilization of ecosystems, loss of biodiversity, soil degradation, desertification, threats to food security, marginalization, rural exodus
Overexploitation Syndrome	Overexploitation of natural ecosystem	Biodiversity loss, climate change, fresh water scarcity, soil erosion, increasing incidence of natural disaster, threats to food security
Rural Exodus Syndrome	Environmental degradation through abandonment of traditional agricultural practices	Loss of ecosystems and species diversity, genetic erosion, eutrophication, acid rain, greenhouse effect, contamination of water bodies and air, freshwater scarcity, soil degradation, marginalization, rural exodus
Intensive Mining Syndrome	Environmental degradation through the extraction of non-renewable resources	Loss of biodiversity, local air pollution, freshwater scarcity, change in runoff, water pollution, soil degradation, creation of contaminated sites, negative effects on health due to pollution
Mass Tourism Syndrome	Development and destruction of nature for recreational purposes	Loss of biodiversity, enhancement of the green effects by air travel, lack of freshwater supply, soil erosion, inadequate disposal of sewage and waste, fragmentation of landscapes by settlements, high consumption of resources.
Military Activities Syndrome	Environmental destruction through war and military action	Loss of biodiversity due to chemical warfare agents, permanent soil degradation due to mining, contamination caused by fuels and explosives, health hazards, greater flows of refugees

Table 2. Patterns of environmental stress: Developmental Syndromes

Development Syndromes	Description	Symptoms
Construction Syndrome	Environmental damage as a result of large construction projects	Loss of biodiversity, local or even global climate change, shortage of fresh water, soil degradation, forced resettlement of local population, danger of international conflicts
Green Revolution Syndrome	The introduction of inappropriate farming methods	Loss of biodiversity, genetic erosion, groundwater pollution, soil degradation threats to food security, health hazards through pesticide use, marginalization, rural exodus, reduction of cultural diversity, reinforcement of regional economic disparities
Asian Tigers Syndrome	Rapid economic growth and disregard for environmental standards	Enhanced greenhouse effect, local climate change, smog, acid rain, water pollution, health hazards, high consumption of resources
Urban Growth / Urban Sprawl Syndrome	Environmental degradation through uncontrolled urban growth /Destruction of landscape through expansion of urban infrastructures	Air pollution, soil erosion, accumulation of waste, noise, population growth, rural exodus, acute health hazards, socioeconomic marginalization, failure of public administration, lack of basic infrastructure, over-oaded traffic infrastructure, fragmentation of ecosystems
Major Accident Syndrome	Unusual anthropogenic environmental disaster with long term impact long term impact	Loss of biodiversity, ecosystem degradation, contamination of soil, water and air, health hazards

The Syndrome Approach identifies different types of interactions, which occur in various environmental, administrative or geopolitical regions of the world. Certain syndromes are predisposed than others to occur due to the onset or escalation of conflict.

Overall importance of the syndrome-based approach is that it may serve as a promising starting point for the development of indicators for early warning and intervention in the conflict dynamic and may provide the opportunity to reduce the potential incidence of conflict or its escalation in specific cases.

Table 3. Patterns of environmental stress: Contamination Syndromes

Contamination Syndromes	Description	Syndromes
Smokestack syndrome	Environmental degradation through large scale diffusion of long lived substances	Loss of biodiversity, eutrophication of ecosystems, depletion of the stratospheric ozone layer, increased levels of UV-B radiation falling on the Earth's surface, enhancement of the greenhouse effect, regional and global climate change, sea-level rise, acid rain, contamination of soils and groundwater with impact on drinking water resources
Waste Dumping Syndrome	Environmental degradation though (un) controlled disposal of waste	Contamination of soils and groundwater, with harmful effects on drinking water, health hazards
Contaminated Land Syndrome	Local contamination of environmental assets at industrial locations	Loss of biodiversity, deposition of pollutants in soils, water and air, loss of soil functions, health hazards
(Modified from WBGU, 1997)		

Table 4. Core problems caused by individual syndromes (Modified from WBGU, 1997)

Core problems / Syndrome	Climate change	Loss of biodiversity	Soil degradation	Scarcity and pollution of freshwater	Threats to world health	Threats to food security	Population growth and distribution	Man-made disasters	Over exploitation and pollution of the world's oceans	Global disparities in development
Overcultivation		x	x	x		x	x	x		x
Overexploitation	x	x	x	x				x	x	x
Rural Exodus		x	x			x	x	x		x
Intensive Mining		x	x	x						
Mass Tourism		x	x	x				x		
Military Activities		x	x		x	x	x			x
Construction	x	x	x	x			x	x		x
Green Revolution		x	x	x	x	x	x			x
Asian Tigers	x	x	x	x	x		x			x
Urban Growth / Urban Sprawl	x	x	x	x	x		x			x
Major Accident		x	x		x					
Smokestack	x	x	x		x	x		x		
Waste dumping		x	x		x					
Contaminated Land		x	x		x					

REFERENCE:

1. German Advisory Council on Global Change (WBGU), World in Transition: The Research Challenge, Springer Verlag, Berlin, 1997

8

RISK ASSESSMENT

HAZARD ASSESSMENT OF MAALAEA CONSTRUCTION AND DEMOLITION LANDFILL SMOLDERING FIRE, AND AIR QUALITY AT A NEARBY COMMUNITY

Jon-Pierre Michaud
Department of Chemistry, University of Hawaii at Hilo. 200 W. Kawili St., Hilo, HI 96720,
E-mail: jonpierr@hawaii.edu

Abstract: This paper presents a chemical monitoring approach taken to assess the hazards and risks to a community posed by a fire in a nearby landfill. Air quality was assessed at the downwind limb of the landfill as well as at the nearest sensitive receptor: the closest community. The hazard assessment included chemical sampling and analysis followed by a toxicological risk-based evaluation of the levels of airborne gasses and aerosols detected. The risk assessment included likelihood of current and future exposures to airborne contaminants in the context of the landfill fire and the surrounding environment. This also included the most notable industrial, agricultural and natural sources of air pollutants in the air shed typically upstream of the community. Concentrations at the landfill were notable while those at the community a mile away were not deemed a health hazard at the time of sampling.

Key words: Landfill fire, hazard assessment, risk assessment, Hawaii community

1.   INTRODUCTION

A Construction and Demolition landfill near Maalaea Maui caught fire on about 26 January, 1998. Landfill operators retained consultants to assist in putting out the fire. After repeated injections of liquid carbon dioxide and application of a soil cap, the fire was subdued to the point of no visible smoke as of early February 1998. Residents of the Maalaea condominium community condominiums, about one mile due south, had complained of obnoxious odors and heath effects ranging from mild to severe. The Hawaii State Department of Health (DOH) sent the first general field investigators to the scene beginning 2 February, followed on 19-21 March 1998 by an engineer sent to assess the status of the landfill, and a chemist/ toxicologist to assess the hazards and risks at the nearby Maalaea community. This paper reports only on the hazard / risk assessment done in March 1998. The aim of that hazard / risk assessment was to evaluate if and to what extent the Maalaea Landfill was a source of risk from hazardous emissions affecting the Maalaea condo community.

2.   METHODS

No water wells were located down gradient from the landfill and significant leaching past the landfill liner was not found; hence no further hydrological assessments were undertaken. Airborne emissions from the landfill were characterized by air samples taken at the landfill and airborne concentrations were also measured at the nearest community.

At the landfill, air was sampled and monitored both at the rim of the landfill on the downwind edge, as well as within the landfill pit itself. Repeated downwind sampling and monitoring transects were made in arcs both on the rim and in the pit (figure 1). This approach allowed an integrative collection of short-lived, transient plumes capturing uneven, heterogeneous distributions of contaminants, which may have been missed by sampling at a single point. This was especially important near the source of contaminants before substantial aerial mixing has had an opportunity to blend the contaminants into a more homogeneous and dispersed plume.

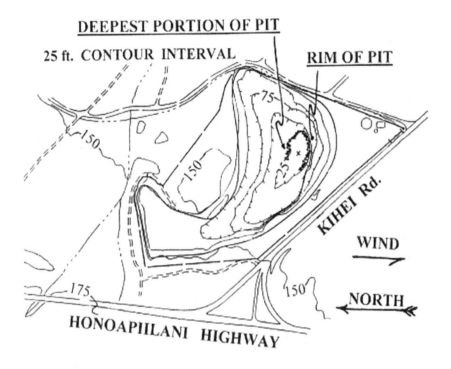

Figure 1. Map of Landfill Area

The first round of the landfill monitoring assessed the site abundance of aerosols / particulate matter less than or equal to ten micrometers in mean aerodynamic diameter (PM_{10}) and particulate matter less than or equal to two and one half micrometer in diameter ($PM_{2.5}$), as well as hydrogen sulfide (H_2S). Nephelometers: an MIE Dataram (DR2000) and personal dataram (pDR-1000) (Monitoring Instruments for the Environment, Bedford, MA) were used to monitor the $PM_{2.5}$ and PM_{10} aerosol loads at the landfill. Aerosols were monitored in real-time along repeated downwind transects in arcs both on the upper most rim and along the rim of the deepest portion of the pit (~ 30m deep, figure 1). Winds were from the north at 5-10 miles per hour. The M1E DR-2000 and the MLE pDR-1000 were both zeroed in filtered air, and both used a 60 second signal averaging period. These monitors were designed as optical surrogates of aerosol mass with limits of detection of 0.1 and 1.0 microgram per cubic meter of air respectively. Seven 10 second near real-time analyses of hydrogen sulfide (H_2S) were made by a Jerome H_2S analyzer (Arizona Instruments, Phoenix, AZ) along seven segments of the previously described downwind sampling tracks (figure 1, both upper and inner arcs). The Jerome H_2S analyzer has a limit of detection of 1 part per billion.

The second round of air sampling and analysis at the landfill (the same day: 3/19/98) was done to assess the abundance of volatile organic carbon (VOC) and reduced sulfur gases down to parts per billion (ppb) levels, as well as benzene, xylene and arsenic trioxide down to parts per million (ppm) levels. Benzene, xylene and arsenic trioxide were sampled within the landfill pit using Drager color change indicator tubes following manufacturer's instructions. VOCs were collected in an integrated sample by an evacuated, passivated stainless-steel "Summa" canister fitted with a 30 minute orifice and 7 micrometer fitted stainless-steel inlet filter (Performance Analytical Inc.). The Summa canister (300A) was allowed to fill for 30 minutes while carried along repeated downwind transects made in arcs on the uppermost rim and along the rim of the deepest portion of the pit (figure 1). Reduced sulfur compounds were sampled for on the same sampling pass, using a "Tedlar" bag (300B). The bag was half filled with air from the sampling arc along the upper rim of the landfill pit, and the remainder of the bag was filled with air from the lower sampling arc along the rim of the landfill. Summa canisters and Tedlar bag air samples were sent by 24-hour express mail, to Performance Analytical Inc. (Osborne, CA) which specializes in air quality analyses. Tedlar bag samples were analyzed for twenty organic sulfur compounds using gas chromatography and flame ionization detection, while summa samples were analyzed for 81 volatile organic compounds and tentatively identified compounds using EPA method TO-14 via cryogenic trapping followed by gas chromatography / mass spectroscopy[1].

At the nearest community (Maalaea), a distance of approximately one mile, it was expected that the small scale (a few meters) heterogeneity would have blended into larger scale (tens to hundreds of meters) Gaussian distributed plumes when the atmosphere is relatively stable. Air in the community was sampled and monitored at the condos closest to the landfill because these sites were expected to give the highest concentrations of any landfill emissions arriving at the community. Subjective odor and symptoms assessments were also made by spending two nights at the condo nearest the landfill.

Air in the community was sampled for VOCs by Summa canister (8-hour orifice, and 7:m stainless steel frit filter) and for reduced sulfur gasses by Tedlar bag. These samples were 24 hour express mailed to Performance Analytical Inc. and analyzed as described above. Aerosols ($PM_{2.5}$ and PM_{10})

and H_2S were monitored as previously described above. Winds were from the north, the direction of the landfill at 2-3 miles per hour.

Seventeen individual resident case histories were taken and symptoms reported over the past month ranged from mild (foul odors) to severely irritating of sinuses, eyes, headache, bleeding nose and peeling skin. Symptoms during the time of the air quality monitoring done in this report ranged from no unusual odors (5 individuals) to noticeable odor with no symptoms (4), to 'it smelled horrible' (2)

3.   RESULTS

At the landfill, an odor reminiscent of a recently burnt and extinguished building was readily apparent in the pit and on the rim of the pit. No obvious sulfide or amine odors were apparent. No smoke was visible.

At the Landfill, aerosols $PM_{2.5}$ and PM_{10} were found as follows. PM_{10} levels (60 second roiling averages) along the upper rim of the landfill ranged from 5 to 35 :g/m^3 and averaged about 10 :g/m^3. Along the lower rim, just above the deepest portion of the pit, PM_{10} levels ranged from 5 to 87 :g/m^3 and averaged about 25 :g/m^3. The overall average PM_{10} level measured for the landfill was 14 :g/m^3. $PM_{2.5}$ levels (60 second rolling averages) along the upper rim of the landfill ranged from 5 to 70 :g/m^3 when including the brief high readings in the mist plume, and averaged about 5 :g/m^3 otherwise for the entire landfill transects. Along the lower rim, just above the deepest portion of the pit, $PM_{2.5}$ levels ranged from 2 to 8 :g/m^3 and averaged about 4 :g/m^3. The overall average $PM_{2.5}$ level measured for the landfill was 7.3 :g/m^3, including the unusually high readings in the water mist plume. That the 2.5 :g/m^3 average in the pit was notably less than the 10 :g/m^3 average in the pit, indicates that much of the aerosol in the pit was larger than 2.5 :g/m^3 and smaller than 10 :g/m^3. This is consistent with dust produced by the truck traffic and is not consistent with smoke aerosols, which tend to be smaller than 2.5 :g/m^3. This suggests the aerosol in the pit was primarily dust and not smoke. This is consistent with the observation that no smoke was visible anywhere in the landfill.   Hydrogen Sulfide (H_2S) was minimal and the highest readings obtained in the landfill were 1 to 2 parts per billion (ppb). Volatile Organic Compounds detected were in the lower middle to low ppb or in the mid to upper parts per trillion. Acetone was the most concentrated at 120 ppb (0.120 ppm). Much higher exposures to acetone would be expected when using nail polish remover. Reduced Sulfur Compounds detected found carbon disulfide at 2.39 parts per billion, not enough to be a health hazard. Carbon monoxide (CO) readings along the upper rim were below detection (limit of detection: 1 ppm). CO readings in the pit were typically 1 to 5 ppm with peaks of 10 to 11 pm. Benzene and Xylene were below the limits of detection (0.5 and I ppm respectively) by colorimetric assay tubes designed specifically for benzene and xylene. Arsenic trioxide was below the limit of detection (5 ppm) by colorimetric assay tubes designed specifically for arsenic trioxide.

At Maalaea Condos, aerosols $PM_{2.5}$ and PM_{10} found were as follows. PM_{10} levels ranged from about 1 to 35 :g/m^3 and averaged 7 :g/m^3. The $PM_{2.5}$ levels ranged from about 1 to 5 :g/m^3 and averaged 2.5 :g/m^3. The only H_2S detected near the condos was found in the immediate vicinity of the septic tank field at one of the condos. Nearby concentrations were only on the order of 1 ppb. Concentrations just over the septic tank lids and at the holes in those lids ranged from 8 to 68 ppb. Of the VOCs found, all compounds detected were in the low ppb or the mid to upper parts per trillion (see appendix 3, sample 325A for data). Of the reduced sulfur compounds, carbon disulfide was detected at 2.07 parts per billion, not enough to be a health hazard.

Other observations found that there were several potential sources of smoke, dust and aerosols in central Maui and that they are typically upwind of the Maalaea Condos (Figure 2). The cane fields in the central area can also be intermittent sources of dust and smoke and the Maui Electric Company (MECO) Maalaea power station, as well as the Kealia pond and wetlands are all within a mile or two and typically upwind of the Maalaea Condos. The MECO smokestack was releasing a visible plume at which, on several occasions in two days and nights, appeared to closely miss the condos along the southeast side, moving to a direct path towards the condos to a near miss to the NW side. An easily audible jet engine like roar seemed to come from the direction of the MECO stacks was heard from the nearest condo rooftop and along the adjacent beach (not quite as audible on beach). Residents have commented on visible dust plumes from the dried portions of the wetland lakebed reaching the community during windy days.

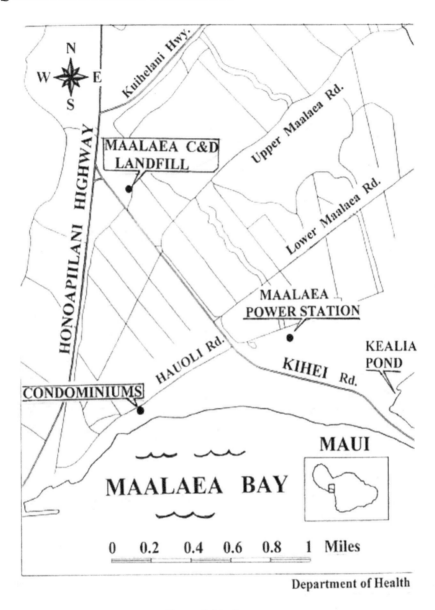

Figure 2. Maalaea Area

Results of the chemical sampling and analyses from both the land fill and the community were compared against the currently available toxicology data bases: the EPA Permissible Remediation Guidelines for chronic exposure, and the Tomes database (Micromedix). Some levels at the landfill were in excess of what was considered 'safe' using available data and guidelines (e.g. $<10^{-6}$ increase risk of cancer) for intermittent and long-term exposures. Levels at the community were below those indicated as causing significant additional risks to health.

4.   CONCLUSION

Based on the chemistry data collected and the toxicology data consulted, there was no clear evidence that emissions from the Maalaea landfill presented acute (short-term) or chronic long-term) health hazards to the Maalaea community.

At landfill proper, none of the compounds detected were expected to present any short-term health hazards until they are up in the low to mid ppm range with the exception of benzene. Benzene was the only compound detected for which a short-term health based guideline (0.1 ppm) below 1.0 ppm could be found. The level of benzene detected at the Maalaea Landfill was 0.089 ppm (89 parts per billion). While these levels are all less than the short-term health based guidelines, they do indicate it would not be recommended to take up residence in the landfill pit.

At the community, none of the compounds detected were expected to present any short-term health hazards until they are in the low to mid ppm range with the exception of benzene. The level of benzene detected at the Maalaea community was 0.00047 ppm (0.470 parts per billion). This is less than 1/212th of the short-term health based guideline of 0.1 ppm.

Because the fire was out and the initial fire was short (weeks), the exposure scenario did not support the use of chronic exposures and lifetime exposure limits. While these limits were considered, they were not used as the primary criteria for assessing risk. Most long-term risk estimates in the literature are based on an increased (additional) risk of 10^{-6} increase in the lifetime risk of cancer. It should be noted that the toxicology literature is always incomplete and does not consider all possible endpoints such as neurological, immunological and reproductive outcomes. Further, there is a paucity of long-term and short-term toxicity data on most of the thousands of compounds in common use, to say nothing of the majority of pyrolysis products resulting from partial combustion such as occurs in smoldering fires. Not all compounds among the most salient of long-term hazards from waste combustion such as dioxins and furans were measured in the course of this investigation. Even though exposures were a single episode, it would have been prudent to have measured levels of these compounds. Further, undetected chemicals may have been present at that or other times which makes it possible for the health hazards to be under-reported and symptoms under-explained.

It should also be noted that there are many sources of the compounds detected. Volatile organic carbon gases can come from burning, from living plants, from decaying organic matter, from gasoline refueling, from car exhaust, from wetlands, and from the ocean to name a few sources. Aerosols can come from burning almost anything, from hydrated oxides of nitrogen or sulfur, from dust lofted by traffic or by working the earth in agriculture, from dried sea spray (tiny salt crystals), and from pollens, mildew and molds. Hydrogen sulfide and reduced sulfur gases can come from sewage treatment, from decomposing organic matter such as in compost piles, municipal solid waste landfills, wetlands, from rubber products and also from volcanoes. Coastal wetlands such as those near the community are often strong sources of these gases. The nearby (and audible) turbine generators at Maui Electric Company (MECO) typically use relatively clean burning natural gas. Because the MECO plant near Maalaea had not added any new generators, had significant changes in operation since 1993, and had not had any Notices of Violation in the preceding six months, the MECO power plant was not likely to be the cause of significant changes in Maalaea community health status during the landfill fire.

The symptoms reported may have had a variety of contributing sources: dust, pollens, molds, smoke, irritant gases and aerosols. Such sources vary with time, come from a variety of both outdoor and indoor sources, and should not be excluded from consideration if symptoms persist. Allergies to specific airborne allergens can be assessed by a skin test by a family practice physician. While not causing damage to cells or tissues, odors, by themselves, can evoke very real symptoms in some individuals. Further site assessment is indicated if obvious symptoms persist or if there are any significant and unfavorable changes in the status of the landfill.

ACKNOWLEDGMENTS

This work was funded by the State of Hawaii and by the Maalaea Construction and Demolition Landfill.

REFERENCES

1. TOMES® System: MICROMEDEX, University of Connecticut Libraries; 369 Fairfield Rd Storrs, CT 06269.
2. EPA method TO-14; Compendium of Methods for Determination of Toxic Organic Compounds in Ambient Air; EPA 600/4-84-041, US Environmental Protection Agency, Research Triangle Park, NC, April 1984, and May 1988.
3. EPA Permissible Remediation Guidelines, Risk Based Screening Levels, 1998.

METHODS FOR ESTIMATING THE IMPACT OF HYPOTHETICAL DAM BREAK FLOODS

Jene Michaud[1], Carl Johnson[2], Judy Iokepa[2], and Jillian Marohnic[2]

[1] *Department of Geology, University of Hawaii at Hilo, 200 W. Kawili St., Hilo, HI USA, 96720, E-mail: jene@hawaii.edu*

Abstract: This paper presents a method for estimating the impact of floods resulting from dam failure. These methods were developed for implementation at the Pacific Disaster Center in Hawaii. A hydraulic model embedded in a Geographic Information System (GIS) is used to estimate the downstream attenuation of the flood. The model output is then integrated with a Digital Elevation Model in order to estimate the lateral extent and depth of flooding. Information useful for disaster planning is obtained by overlaying the flood map onto GIS infrastructure layers (roads, emergency services, and chemical plants).

Key words: GIS, Floods, hazard assessment, risk assessment, dam breach, Hawaii

1.   INTRODUCTION

In many locations around the world dams are essential for agriculture and urban water supply. While the benefits of these structures are obvious, the risk of damage or loss of life due to unexpected failure of the dam can be overlooked by the general populace. A systematic assessment of the hazards posed by possible dam failures is useful for disaster planning, emergency response, and flood warnings.

A hypothetical scenario will be used to orient the reader to the problem. A town by a river lies 20 km downstream of a tall dam. What will happen if the earthen dam suddenly crumbles in an earthquake? Will the water in the city streets be 0.5 m. deep or 5 m. deep? Will the water be swift enough to erode roads and sweep people away? Will floods affect chemical plants, landfills, water treatment plants, and stores of hazardous materials? Are lives at risk, and if so, how many? How long will it take the flood to travel from the dam to the town? How long will the flooding last? These questions are addressed by the methods presented below.

Because there is no regulatory demand for dam breach risk assessments, there are no standardized methods of analysis. This paper presents the methods developed by the authors for implementation at the Pacific Disaster Center on Maui. The resulting protocols and software programs are intended for application to high hazard dams in Hawaii and possibly other locations in the Pacific. Some of the GIS-based methods presented below are applicable to any hazard that occurs over a broad area.

2.   METHODS

As a practical matter, it is more productive to estimate the consequences that are expected from a standardized dam breach scenario than to examine a range of possibilities in order to estimate the expected annual damages. Damages will vary with the volume of water in the reservoir. For general planning purposes the most useful scenario is the worst-case one in which the dam fails while the reservoir is full.

We have established a protocol for estimating the impacts of specific dam breach scenarios in Hawaii. There are three major steps: conducting one-dimensional hydraulic analyses, producing two-dimensional flood maps, and estimating the impacts on the community.

Step one consists of modeling the hydraulic characteristics of the flood, including the discharge, speed, elevation, and depth of the flood at various times and at various distances downstream of the dam. This step is accomplished using a hydraulic model called FLDWAV, which was developed specifically for dam breach situations by the U.S. National Weather Service (NWS).[1] FLDWAV models the breach of the dam and emptying of the reservoir, and then routes the resulting flood downstream. The routing problem is a particularly difficult one due to extreme changes in hydraulic conditions over a short time. FLDWAV is considered to be the state-of-the-art in model for dam breach floods, but even experienced modelers find it difficult to use. Some users therefore prefer to use the less accurate — but more user friendly – "Simple Dambreak" model developed by the NWS for preliminary analyses.

The hydraulic models require topographic information in the form of the stream's longitudinal profile and channel cross-sections at several locations.

We have developed GIS-based procedures for extracting the needed information from a GIS-based Digital Elevation Model (DEM). The DEM data can also be supplemented with precise ground surveying of the channel. The GIS software program that we have developed (an ArcView application called "M2M") coordinates the extraction of topographic data, running of the FLDWAV model, and importing the FLDWAV output back into the GIS.

FLDWAV output is one-dimensional, and must be convolved with a two-dimensional topographic map (the DEM) in order to estimate the lateral extent of flooding and to produce a map showing the spatial variations in flood depths. This is accomplished using an "onion skin" technique for spreading the flood laterally in small increments until it reaches a topographic barrier. A transmission loss option prevents the flood from spreading to infinity on the flat coastal plains. The final result is a map that depicts spatially variable flood depth using various shades of blue. The water velocity (averaged across the cross-section) can also be depicted on the map.

A first-order estimate of the impacts of the flood scenario can be obtained through an evaluation of the flood depth map after it has been overlaid onto an electronic version of a standard topographic map that depicts roads, buildings, and public facilities. Such maps are useful for preparing evacuation plans and selecting locations for roadblocks. The FLDWAV output files may be consulted to determine the time characteristics of the flood, such as the time from the dam breach to the onset of flooding, and the duration of flooding.

A more formal impact analysis can be obtained by overlaying the flood inundation maps onto GIS infrastructure layers such as roads, schools, and chemical plants. This is discussed in more detail in the case study presented in section three. While not attempted in this study, it is possible to expand the GIS analysis to estimate the direct economic losses due to flooding. (Indirect losses such as lost time at work, reduced business' activity, and clean up of flood-related contamination can also be significant, but are more difficult to estimate.) Estimation of direct losses could be made by using GIS-based tax maps that are linked to computerized tax assessments. The value of losses would be estimated by applying standardized loss functions that are based on building type and flood depth. Similar procedures are being developed within the HAZUS model to estimate losses from the 100-year flood. HAZUS is a software program developed by U.S. Federal Emergency Management Agency to estimate potential losses due to natural hazards. [2]

3.   HAWAIIAN CASE STUDIES

The methods described above were tested at two sites in Hawaii: The Nuuanu reservoir on Oahu, which is above downtown Honolulu, and the Waikoloa Dam on Hawaii Island, which is above the town of Waimea. In both cases the analyses were performed with and without topographic data obtained by a field survey crew. Detailed results from the case studies and results of a sensitivity analysis are reported elsewhere.³ The flood inundation maps produced for Waimea and Honolulu were overlaid onto several GIS infrastructure layers.   These layers included major roads, secondary roads, schools, nursing homes, hospitals, police stations, fire stations, civil defense headquarters, chemical plants, electric plants and transmission lines, water plants, and wells (which could be contaminated by floodwaters). Critical facilities in the flood zone were identified and listed along with their mailing addresses and phone numbers of contact personnel.

Part of the purpose of the case studies was to gain experience that would guide future dam break analyses in Hawaii. Some of the lessons learned during the case studies are therefore listed below. These conclusions are based on the two sites examined but cannot automatically be assumed to hold for all other sites.

a) The DEM does not adequately represent the inner stream channel. This deficiency can be remedied with a simple field survey of the channel cross-section, but it is unclear whether the additional cost is worthwhile

b) Because the final flood maps have the potential to affect land values, some landowners do not wish to facilitate surveying of the channels on their property. It appears that the potential legal liability that comes with making specific hazard assessments must be taken seriously. This is a particular concern because formal evaluation of the errors in the flood hazard assessment would be exceedingly difficult. The decision of whether or not to release dam breach flood maps to the general public will be a difficult one.

c) Sensitivity studies at the test sites suggest that the estimated first-order impacts are not particularly sensitive to the selection of parameter estimates or to the choice of topographic data. In other words, the main conclusion (for example, a flood of up to 1-2 meters will sweep through the commercial district within 30 minutes of the dam failure) appears to be relatively robust even if there are uncertainties in the exact depth, velocity, and lateral extent of the floodwaters.

d) The impact assessment is only as accurate as the GIS infrastructure layers. At one of the test sites it was observed that some existing facilities (a school, a landfill) were not represented in the current generation of GIS data.

4.   CONCLUSION

The protocols and software developed in this study will give the Pacific Disaster Center tools that will enable it--and other agencies such as the Hawaii State Department of Land and Natural Resources and the U.S. National Weather Service--to conduct dam breach hazard assessments more easily. Results of these assessments are expected to be of value for disaster planning, emergency response, and flood warning.

ACKNOWLEDGMENTS

This work was funded by the U.S. National Aeronautical and Space Administration (Contract NASW-99044). The DEM and GIS infrastructure data were supplied by the Pacific Disaster Center. The method for preparing the two-dimensional flood maps is based in part by a method described to us by Andy Rost of the National Operational Hydrologic Remote Sensing Center.

REFERENCES

1. Fread, D., Lewis, J., FLDWAV: A generalized flood routing model, in: Proc. of National Conference on Hydraulic Engineering, eds. Abt S., Gessler, J, American Society of Civil Engineers, 1988, pp. 668-673, ISBN 0872626709.
2. United States Federal Emergency Management Agency, About Hazus, http://www.fema.gov/hazus/ab_main.htm, June 26, 2002, <July 1, 2002>.
3. Johnson, C. and Michaud, J., Dam Failure Inundation Mapping Project: Final Report and User Guide (NASA Contract NAWS-99-044), October 2001

8

WASTEWATER AND SLUDGE MANAGEMENT

POLYCHLORINATED DIBENZO-P-DIOXINS (PCDDS) AND DIBENZOFURANS (PCDFS) IN SEWAGE AND SLUDGE OF MWTP

Marzenna R. Dudzińska
Institute of Environmental Protection Engineering, Technical University of Lublin, Nadbystrzycka 40B, 20-618 Lublin, Poland, E-mail: mardudz@fenix.pol.lublin.pl

Abstract: Polychlorinated dibenzodioxins (PCDDs) and dibenzofurans (PCDFs), commonly known as "dioxins", in contrast to other chlorinated chemicals, have never been commercially manufactured or are of any benefit or known use. They enter the environment not as primary products but unintentional trace impurities. PCDD/Fs were found in the emissions of various combustion processes, independently from the fuel, so that thermal processes are recognized as a main source, especially processes of waste incineration. But "dioxin problem" is strictly combined with all human activities connected either with solid wastes or wastewaters production and utilization.

CDD/Fs presence in municipal wastewater treatment plant (MWTP) influents results in the PCDD/Fs presence in effluents and sludge. Evidence of polychlorinated dibenzo-p-dioxins and dibenzofurans formation within the MWTP was also found. All this aspects are discussed in the paper, with special attention to sewage sludge, because of further utilization demand according to EU directives.

Key words: Polychlorinated dibenzo-p-dioxins (PCDDs); polychlorinated dibenzo-p-furans (PCDFs); sewage sludge; household sewage; and municipal wastewater treatment plant (MWTP)

1.   INTRODUCTION

The growing amount of sludge, generated every year by wastewater treatment plants is one of the crucial environmental problems. From one side, improved methods of wastewater treatment result in the growing amounts of sludge, from other, the minimization of waste became a "prime directive" of modern societies.

Growing amounts of sludge require proper way of treatment and utilization. Existed and applied methods of utilization, like incineration or deposition, have advantages and disadvantages. Incineration is believed to be a method of total destruction, although some residues remain and hazardous emissions require expensive methods of reduction and monitoring. Landfilling is cheaper, but lack of new territories and leakages into surface and groundwaters are "hot issues". According to European Community data, most of sludge in Europe, till the middle 90-ties was deposited (including 6% deposited in the sea), 11% was incinerated, and 37% was reuse in agriculture[1]. Since 1995, deposition in the sea has been banned. Additionally, directives prohibiting deposition of any waste which may be reused have been introduced. So the interest in the agriculture utilization is growing, although a lot of limitations have to be considered, as sewage sludge is the well know abiotic sink for persistent chemicals. Pathways of heavy metals have been studied for years, and strict regulations introduced in many countries. The environmental pathways of stable, hydrophobic chloro-organic pollutants, which may easily accumulate in the food chain, have not been fully recognized, as is the evidence of enzymatic reactions leading to formation of some hazardous compounds during natural composting processes and thermal sludge treatment. Among organic compounds, persistent in either aerobic or anaerobic conditions, polychlorinated dibenzo-p-dioxins and dibenzofurans (PCDD/Fs) are the most hazardous for human beings.

2.   FEW FACTS ABOUT PCDDS AND PCDFS

Polychlorinated dibenzodioxins and dibenzofurans (PCDD/Fs), in contrast to other chlorinated chemicals, have never been commercially manufactured nor are of any benefit or known use. PCDD/Fs are ubiquitous contaminants, which are released as byproducts of incomplete combustion or as impurities in chemical processes, and that their levels in the environment are increasing.

The name of PCDD/Fs or dioxins is given to a family of compounds of different number of chlorine atoms in the molecules (congeners). Totally, there are 75 of polychlorodibenzo-p-dioxins and 135 polychlorodibenzo-p-furans that might be formed. Almost all of them have been found in the environmental matrices (air, soil, sediments and animals tissue) and abiotic reservoirs, but only 17 congeners, considered to be of the highest toxicity are usually measured. The PCDD/Fs level is presented as TEQ, e.g. sum of 17 congeners levels multiplied by their toxicity factors (TEF), and only this value is regulated by law in many countries. The highest toxicity factor TEF = 1 was attributed to 2,3,7,8-tetrachlorodibenzo-p-dioxin (2,3,7,8,-TCDD), which in some animals exhibited extreme toxicity (the lethal dose for 50% mortality (LD_{50}) for 2,3,7,8-TCDD, is 45 µg/kg in female rats and 1 µg/kg in guinea pigs, respectively) [2].

Furthermore, all 2,3,7,8-substituted compounds persist in the environment and have been found to strongly bioaccumulate in fish and mammals. Measurement and calculation of TEQ is considered to be sufficient for estimation of PCDD/Fs hazard to the humans and environment. In 1999 European Community, after intensive studies and WHO recommendations introduced new, revised TEFs, so comparison of recent results with previously published in literature needs special attention [2].

Every dioxin source has different isomeric pattern (different ratio of concentration of individual isomer within all group of isomers) and congener profile (total concentration of one isomer group compare with other groups of isomers). Congener profiles and isomeric patterns might be used to determine the main source in different matrices, but congeners differ not only in toxicity, but in physico-chemical properties, like solubility in water, resistance to photodegradation and photodechlorination, sorption, that are crucial to their behavior in different matrices. So also the matrix itself might influence the pattern [3].

The primary sources of PCDDs and PCDFs in the environment can be divided into four categories: chemical reactions, thermal reactions, photochemical reactions and enzymatic reactions [4].

Chemical reactions, where PCDD/Fs are formed from their precursors in industrial syntheses, resulted in contamination of industrial products like chlorophenols, chlorophenoxy herbicides, pesticides and PCBs [2]. Production and use of these chemicals are now-a-days banned or strictly regulated, but in 1960s and 1970s these widely used products were a major source of PCDD/Fs in the environment. Other chemical processes generating PCDD/Fs result from the paper pulp bleaching, dry cleaning distillation residues and chlorine gas production using graphite electrodes.

Thermal reactions leading to the PCDD/F emissions are connected with technological and domestic combustion. PCDD/Fs were found in the emissions of the various combustion processes independently from the fuel: municipal and hazardous waste incinerators, power plants with fossil fuels, automobile exhaust, private heating and fire places, wood and forest fires, cigarette smoking as well as accidents like fires of polyvinyl chloride (PVC) or polychlorinated biphenyl's (PCB) [3]. Copper smelters and steel production also belong to the thermal sources.

Photochemical reactions can result in the formation as well as the degradation of PCDD/Fs. These reactions are carefully examined as most combustion and incineration sources produce emission directly into the atmosphere and they undergo long-distant transport.

According to the recent evidence, enzymatic reactions leading to PCDF/D formation, might occur under true environmental conditions in sewage sludge and during composting processes [5].

PCDD/Fs are emitted or spilled from many processes, but as they were found in the emissions from all combustion processes, thermal processes are recognized to be a main source, and a major public concern, especially waste incineration. But "dioxins problem" is attributed to all human activities, connected either with production or utilization of solid wastes and wastewaters. One of the crucial examples of the problem is sewage sludge contamination.

3.   PCDD/FS IN SEWAGE AND SLUDGE

Sewage sludge contains trace amounts of PCDD/Fs and PCBs and it is therefore important that the environmental implications of this disposal method are fully understood. Concerns center on the possible transfer of these compounds from sludge into the human food chains, mainly via grazing livestock [6]. Attempts have been made to identify and quantify sources of PCDD/Fs in sewage sludge [7].

Taking into account the known sources in the environment, PCDD/Fs presence in the sludge may originate from:
– atmospheric deposition,
– formation from precursors during anaerobic and aerobic treatment,
– sewage itself.
And consequently, dioxin might be introduced into sewages from:
– industrial streams,
– atmospheric deposition, (street run-offs),
– domestic sewage (chloro-organic compounds in textiles, paper, etc.).

Gihr and co-workers [8] estimated PCDD/F sources to sewage sludge and made two important observations. Firstly, known sources of PCDD/Fs (like polychlorophenols - PCP) cannot explain the total burden in sewage sludge and secondly, that while PCP-associated PCDD/Fs can account for the presence of several key congeners, the broad and relatively high background contamination observed cannot be wholly attributed to PCP use. They therefore concluded that the most likely source of PCDD/Fs in sludge was atmospheric deposition onto roads, followed by runoff. In contrast Horstmann and MacLachlan [7] identified household wastewater as more important source of PCDD/Fs than runoff, implicating laundry wastewater as a major source in household wastewater. Naf *et al.*[9] conducted a PCDD/F mass balance for sewage treatment plants but were unable to account for all the load observed. Oberg *et al.*[10] and Oberg and Rappe[5] have reported the "*de novo*" formation of ^{13}C-PCDDs from ^{13}C-PCP in municipal sewage sludge and concluded that biological activity was the cause of this transformation; this may partially explain the discrepancy between known sources and the observed burden in sludge. It may therefore be appropriate to consider sewage sludge as a primary as well as secondary source of PCDD/Fs.

Recently, the contamination of PCDDs and PCDFs in sewage sludge is well documented and analytical results have been reported for USA[11], Canada[12] and UE countries[1]. However, there are no data for a number of European countries, and Poland is one of them. So far, in Poland only levels in emissions from selected incineration processes and levels in food samples have been measured [13]. There is a crucial need to estimate PCDD/Fs loads from sources in Poland, including sewage sludge utilization.

4.   PRELIMIARY MEASUREMENTS OF PCDD/F LEVELS IN POLISH SLUDGE

First attempts to measure the PCDD/F levels in Polish sludge were made in 2001 for sludge samples from Municipal Wastewater Treatment Plant in South-Eastern Poland [14]. "Hajdow" Municipal Wastewater Treatment Plant collects municipal wastewater discharge by sewage systems from Lublin, city of 400 thousands inhabitants and Swidnik, small town of 20 thousands inhabitants. This part of Poland is generally agriculture, with relatively small industrial enterprises (food and machine). Helicopter factory located in Swidnik has separate wastewater treatment system. Industrial streams from Lublin (brewery, beet-sugar factory, and pharmaceutical plant) are discharged to municipal sewage system after primary on-site pretreatment. Increasingly intensive car traffic from the West to Ukrainian and Belarusian borders have been passing through Lublin in the last few years, but street run-offs are collected by separate system and discharged directly to the surface waters.

"Hajdow" MWTP is a typical Polish municipal wastewater treatment plant and consists of primary (mechanical) treatment part and secondary (biological) treatment part. Total treatment procedure lasts about 22 hours. The primary sludge and excess sludge are water reduced separately and go together to the digester.

Samples were collected in spring 2001 (March and April) during dry weather period. Sample preparation and measurement procedures were described elsewhere[14]. In autumn, which is the "campaign time" for different food processing factories (mainly sugar-beet factory) the loads might differ. Obtained results are gathered in Table 1.

Table 1. PCDD/F concentrations (ng/kg dry matter) in the sludge samples.

Congener	S/04/01/1	S/04/01/2	S/03/01/2
2,3,7,8-TCDD	1.1	ND	1.3
1,2,3,7,8-PeCDD	4.89	5.62	2.1
1,2,3,4,7,8-HxCDD	6.35	6.38	4.5
1,2,3,6,7,8-HxCDD	234	119	68
1,2,3,7,8,9-HxCDD	35.8	32.6	24
1,2,3,4,6,7,8-HpCDD	2388	2838	1900
OCDD	12800	14700	9200
2,3,7,8-TCDF	5.12	4.17	18
1,2,3,7,8-PeCDF	4.72	3.72	6.95
2,3,4,7,8-PeCDF	5.24	4.94	9.66
1,2,3,4,7,8-HxCDF	6.32	6.32	8.5
1,2,3,6,7,8-HxCDF	6.95	5.75	11
1,2,3,7,8,9-HxCDF	ND	ND	ND
2,3,4,6,7,8-HxCDF	7.34	8.34	12
1,2,3,4,6,7,8-HpCDF	96.50	56.5	210
1,2,3,4,7,8,9-HpCDF	8.25	6.25	9.87
OCDF	234	185	780
PCDD/Fs WHO-TEQ	**65.26**	**57.02**	**53.50**
PCDD/Fs I-TEQ	**74.55**	**67.62**	**63.05**

ND – not detected

TEQ calculations were made based on both - the WHO-TEFs and I-TEFs for seventeen 2,3,7,8-substituted PCDDs and PCDFs. "Dioxin like" PCBs were not measured. Total WHO-TEQ PCDD/F values for the analyzed samples range from 53.50 to 65.26 ng/kg dry mass. Calculations, based on I-TEFs were also made, to allow comparison with the literature data. The variation in the TEQ concentration may be attributed to varying loads during the sample collection period. Obtained results are higher, than medium

European value of 20 ng/kg, typical for medium size cities without heavy industry. But still lower than the highest values, around 200 ng/kg, typical of highly industrialized areas, where impact of air deposition and "incineration born" tetra- congeners is high.

Despite of the sampling, the concentrations of PCDDs were higher than those of PCDFs. A general increase of concentration with the increasing degree of chlorination was observed and OCDD was the predominant congener, which is the trend reported by other researchers [15, 16, 17] for urban sewage sludge.

But also tetra-congeners, including tetra-furans have impact for the total TEQ. This may express the influence of atmospheric deposition within the wastewater treatment plant. At this stage of research and lack of archival samples, it was impossible to compare recent levels with historical ones, so any comments about decreasing or increasing concentrations are not justified.

The general increase in concentration with increasing degree of chlorination can be observed, with OCDD (octachlorodibenzo-p-dioxin) as the predominant congener. This is consistent with previously reported data for sewage sludge in other countries.

Table 2. PCDD/F range in sewage sludge from different European countries.

Country	Levels ng I-TEQ/ kg d.m.	Average ng I-TEQ/ kg d.m
Austria	8 - 40	15
Denmark	>1 - 55	10-20
Germany	0.7 - 1200	10-60
Spain	7-160	55 -64
Sweden	> 1 - 115	26
U.K.	9 - 192	20-80
EU	9 - 144	
Poland	50 ; 70	

5.   CONCLUSIONS

Because PCDD/Fs became to be recognized as the most hazardous for the environment and human beings, very restricted emission limits were introduced in the most countries. But only a few countries limits are also pursuit for PCDD/F levels in sewage sludge, however sludge is one of the ultimate sinks for persistent chemicals.

Germany is the country where government has taken far reaching legislative action to reduce PCDD/Fs release into the environment. No restrictions are applied to soils containing up to 5 ng ΣTEQ/ kg in Germany. Certain restrictions apply when the soil TEQ value falls in the range of 5-40 ng/kg, including recommendations to monitor potential increased transfer into the food chain. Above 40 ng/kg there are restrictions on the uses of the soil for agriculture. In July 1992, a limit value of 100 ng TEQ/kg dry matter was established for PCDD/Fs in sludges used in German agriculture. This limit value was set in conjunction with a sludge application rate limit of 5 tonnes of sludge (DW) per hectare over a 3-year period[18]. The situation in Poland is different, as sewage sludge has been used for agricultural purposes in very limited amounts, up till now.

Table 3. PCDD/Fs release to land resulting from sludge application

Country	Sludge production	Sludge used in agriculture	PCDD/Fs (g TEQ) release to land	
	Mg dry solids/year		min	max
Austria	no data	21 900	no data	no data
Belgium	59 200	63 010	0.2	3.2
Denmark	170 300		0.6	9.1
Finland	no data	511 200	no data	no data
France	852 000	730 000	4.6	74
Germany	2 681 299	4 820	6.6	106
Greece	48 200	3 670	0.04	0.7
Ireland	36 700	301 920[a]	0.03	0.5
Italy	816 000	2 920[a]	2.7	43
Luxembourg	7 900	119 470[a]	0.03	0.4
Netherlands	322 900	9 250[a]	1.1	17
Portugal	25 000	129 500	0.08	1.3
Spain	350 000		1.2	19
Sweden	no data	465 000	no data	no data
UK	1 107 000	2 260 000	4.2	67
Total	6 480 000	12 750	21	340
Poland	255 000		0.82	?

[a] based on the EU average

Because of development of the new municipal sewage treatment plants and improving efficiency of the existing ones, it is very difficult to get recent data about annual production of sewage sludge. But in the year 2000, about 1243 million m^3 of sewage was treated in Poland[19], what gives 255 thousands tons dry matter sludge per year, assuming sludge unit index equal 0.217 kg d.m. /m^3 (Polish average [20]). About 52% of sludge in Poland, is disposed on landfills, 33% used in land leveling and remediation of post-industrial degraded area. Only 3-5 % of total sludge production is used in agriculture[20]. We had no possibility to compare our results to concentrations in sludge from the others MWTP in Poland. Assuming roughly, that obtained results may be taken as Polish average, it does not mean "high" annual load of PCDD/Fs to the soil from that source. But to estimate real loads, more measurements are needed. Specially, that growing amount of sludge will force agricultural utilization. And potential risk from utilization in land remediation as well as from deposition on landfills should be taken into consideration.

With regard to aspects of minimization of contaminants and meeting legal requirements limit values in environmental related matrices, the degradation and formation of PCDDs and PCDFs by microorganisms are of special interest. At present the agricultural application of sewage sludge, which is supposed to be the most important accumulation medium for organochlorine compounds, is under discussion because of the persistence of the PCDD/F compounds in soil.

Sewage sludge application to agricultural land is currently under scrutiny to assess it's potential of increasing human exposure to various organics. There are limited field data on the effect of sludge application on levels of PCDD/Fs in the food chain.

The new Polish Waste Act (issued May 2001) strictly regulates sludge application. The disposal of sludge to the land used for vegetable cultivation as well as to pastures and meadows is prohibited, what minimizes the possibility for persistent chemicals to enter the food chain.

Decreasing levels in sludge has been reported in several European countries since nineties[16,17,18]. But we can expect the growing amounts of sludge production after introducing the European Community Directive (COM 91/271) on wastewater treatment, requires the installation of treatment systems in all towns with over 2000 inhabitants before 2005. The need to monitor PCDD/Fs loads from this source remains.

Particular attention has been focused on the application of sewage sludge to agricultural land in the context of PCDD/F additions, given the propensity for the 2,3,7,8-substituted compounds to bioaccumulate into livestock.

Further detailed studies are required to assess the potential transfer of PCDD/Fs (and other organochlorines) to grazing animals under various agricultural management settings, to enable a fuller evaluation of these issues.

The environment is a dynamic system and toxic chemicals inadvertently released into one phase of the environment invariably translocate into the other phases. The potential risks associated with the release of toxic chemicals in the environment are directly related to the extent and rate of translocation of chemicals from one phase into another.

ACKNOWLEDGMENTS

This research was carried out within the Project No 3 T09C 00618, financed by Polish State Committee for Scientific Research.

REFERENCES

1. *European Dioxin Inventory*, 1999, Report for European Commission, DG XI. Hutzinger O., Fiedler, H., Sources and Emissions of PCDD/F, Chemosphere 18, 1989, 23-32.
2. Dudzinska, M.R., Kozak, Z., *Polychlorinated dibenzo-p-dioxins and dibenzofurans – properties and environmental pathways* (in Polish), Monographies of Polish Academy of Sciences, 2001, ISBN 83-915874-4-4.
3. Rappe, C., Sources of PCDDs and PCDFs, introduction, reaction, levels, patterns, profiles and trends, Chemosphere 25, 1992, 41-44.
4. Rappe, C., Sources of exposure, environmental concentrations and exposure assessment of PCDDs and PCDFs, *Chemosphere*, 27, 1993, 211-225.
5. Oberg, L.G., Rappe, C., Biochemical formation of PCDD/F from chlorophenols, *Chemosphere* 25, 1992, 49-52.
6. McLachlan, M.S., Horstmann, M., Hinkel, M., Polychlorinated dibenzo-p-dioxins and dibenzofurans in sewage sludge: sources and fate following sludge application to land, *Sci. Total Environ.*, 185, 1996, 109-123.
7. Horstmann, M. and McLachlan, M.S., Concentrations of PCDD and PCDF in urban runoff and household wastewaters, *Chemosphere* 31, 1995, 2887-2896.
8. Gihr, R., Klopffer, W., Rippen, G. and Partscht, I., Investigations of potential sources of polychlorinated dibenzo-p-dioxins and dibenzofurans in sewage sludges, *Chemosphere*, 23, 1991, 1653-1659.
9. Naf, C., Broman, D., Ishag, R., Zebuhr, Y., PCDDs and PCDFs in water, sludge and air samples from various levles in a waste water treatment plant with respect to composition changes and total flux, *Chemosphere* 20, 1990, 1503-1510.
10. Oberg, L.G., Anderson, R., Rappe, C., De novo formation of hepta- and octa-chlorodibenzo-p-dioxins from pentachlorophenol in municipal sewage, *Organohalogen Compounds*, 9, 1992, 351-354.
11. Telliard, W.A., McCarty, H.B., King, J.R. and Hoffman, J.B., USEPA national sewage sludge survey for polychlorinated dibenzo-p-dioxins and polychlorinated dibenzofurans, *Organohalogen Compounds* 2, 1990, 307-310.
12. Ho, A.K. and Clement, R.E., Chlorinated dioxins/furans in seage and sludge of municipal water pollution plants, *Chemosphere* 20, 1990, 1549-1552.
13. Grochowalski, A., Chrzaszcz, R., The results of the large scale determination of, PCDFsand coplanar PCBs in Polish food product samples using GC-MS/MS technique, *Organohalogen Compounds* 47, 2000, 310-313.

14. Dudzinska, M.R., Czerwinski, J., PCDD/F levels in sewage sludge from MWTP in South-Eastern Poland, *Organohalogen Compounds*, 57, 2002, 305-308.
15. Hagenmaier, H., Linding, C., She, J., Correlation of environmental occurrence of polychlorinated dibenzo-p-dioxins and dibenzofurans with possible sources, *Chemosphere* 29, 1994, 2163-2174.
16. Rappe, C., Anderson, R., Karlaganis, G. and Bonjour, R., PCDDs and PCDFsin samples of sewage sludge from various areas in Switzerland, *Organohalogen Compounds*, 20, 1994, 79-83.
17. Eljarrat, E., Caixach, J., Riviera J., Decline in PCDDand PCDF levels in sewage sludges from Catalonia (Spain), *Environ. Sci. Technol.*, 33, 1999, 2493-2498.
18. Sewart, A., Harrad, S.J., McLachlan, M.S., McGrath, S.P. and Jones, K.C., PCDD/Fs and non-o-PCBs in digested UK sewage sludges, *Chemosphere* 30, 1995, 51-67.
19. *Statistical Yearbook of the Republic of Poland*, 2001.
20. Bernacka, J., Pawlowska, L., *Sewage Sludge Management in Poland (in Polish)*, 1998, ISBN 83-908960-1-X.

THE UTILIZATION OF SLUDGE GENERATED IN COAGULATION OF FISH PROCESSING WASTEWATER FOR LIQUID FEED PRODUCTION

K. Medrzycka, R. Tomczak-Wandzel
Gdańsk University of Technology, Chemical Faculty, ul. Narutowicza 11/12, 81-952 Gdańsk, Poland

Abstract: The offals from fish processing contain substances can have high nutrition value and can be used for feed production. The created fodder could be dry (e.g., fish meal) or liquid (protein hydrolyzates). The nutritive components of fish processing plant wastewater have not been utilized in Poland. A coagulation process, using sulfuric and phosphoric acid, has been developed and optimized to concentrate the proteins and fats in wastewater sludge.

Key words: Fish processing wastewater, post-coagulation sludge, fish silage.

1. INTRODUCTION

In fish processing plants, a huge amount of wastewater is generated. The main components of this wastewater are proteins and fats, substances of nutritious value.

In Table 1, the loads of particular pollutants discharged to wastewater from production of various processes of fish products are presented.

Table 1. The loads of pollutants discharged to wastewater from production of 1 tonne of fish products [1]

Product	Pollutants [kg/t]				
	BOD$_5$	COD	Fat	Suspended solids	Protein
Canned fish	62.0	123.6	39.8	78.6	32.9
Pickles	21.3	74.7	0.3	22.4	25.6
Fish oil	41.1	210.5	61.0	95.0	5.6
Salted fish	56.0	-	3.6	140.0	38.1
Smoked fish	35.0	-	4.0	3.0	10.0
Fish meal	280.0	320	168.0	200.0	258.4

As it can be seen from Table 1 the biggest loses of protein and fat are observed in fish oil and fish meal production. In wastewater the concentration of protein and fat can reach 2.6% and 4%, respectively.

The annual discharge of protein and fat with wastewater from fish processing sector in Poland is presented in Table 2.

Table 2. Protein and fat annual discharge with wastewater from fish processing industry in Poland

Products	Production [t]	Discharge to wastewater	
		Protein [t]	Fat [t]
Canned fish	37580	1236	1495
Pickles	26711	684	8
Salted fish	23612	899	85
Smoked fish	32233	322	129
Fish meal	7543	1946	1267
	total	5087	2984

Such tremendous losses of valuable substances should be avoided. A possibility is to recover them during wastewater treatment [2-4]. The aim of this research was to recover proteins and fats from wastewater in form useful for animal feed production.

Wastewater from fish processing can be treated in different ways: mechanically, chemically or biologically [5]. Chemical treatment it is a coagulation process in which the added chemicals (coagulants) form flocks to which dispersed pollutants easily attach [6]. This material is separated from wastewater by sedimentation. Coagulants, such as aluminum or iron salts are usually used. Post-coagulation sediment from fish processing wastewater consist of coagulant (in form of aluminum or ferric hydroxide) and separated pollutants; e.g., proteins, fats and other colloids. However the main component in these sludge is water. Post-coagulation sludge containing nutritious substances can be used as an additive to animal feed production.

There **are** two possibilities: dry feed – e.g., fish meal and liquid feed – e.g., fish silage (protein hydrolyzates) [7]. In this research the utilization of sludge for fish silage production has been investigated.

Fish silage is produced in a way presented on these scheme [8]:

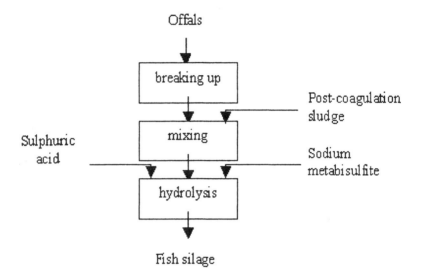

Figure 1. Scheme of fish hydrolyzate production

Fish silage is usually produced from fish offals (heads, intestines, skin, etc.) or waste fish. The enzymes present in fish viscera decompose these tissues and the liquid protein hydrolyzate is the final product [9-11]. To avoid bacteria growth a pH about 2 is required and a relevant amount of organic or inorganic acid is added, usually it is sulphuric acid. The idea was to supply a certain amount of post-coagulation sludge for silage production [12].

2. EXPERIMENTAL

In our experiments the coagulation of fish processing wastewater was performed using ferric sulfate. The sludge generated in this process has been added for silage production. The sludge characteristics are presented in Table 3.

The total amount of iron in wastewater during coagulation depends mainly on the amount of coagulant added. However, as it can be seen from Tables 3 and 4, about 90% of the iron present in raw wastewater and added as a coagulant, is transferred to the post-coagulation sludge. In the sludge its concentration was about 2.6 g/dm^3 (about 0.26 %) which is about 2.6% of dry matter of the sludge (Table 4).

Table 3. Iron concentration in wastewater and in post-coagulation sludge

Process	Iron in raw wastewater [mg/dm^3]	Iron added for coagulation process [mg/dm^3]	Iron in waste-water after coagulation [mg/dm^3]	Iron transferred to post-coagulation sludge [%]
1	17.0	175.0	16.1	91.6
2	19.3	175.0	20.2	89.6
3	23.1	100.0	13.7	88.9
4	23.4	50.0	5.1	93.0

Table 4. Characteristics of post-coagulation sludge

Parameters	[%]	[mg/dm^3]
Water	89.7	-
Fat	6.2	-
Protein	2.7	-
Ash	0.7	-
Total iron	0.26	2599.7
Iron absorbable	0.03	269.5

When the post-coagulation sludge is added to hydrolyzate production, then the iron hydroxide is converted into ferric sulfate, which is well soluble in water. Too high content of soluble iron compounds, which are easily absorbable, should be avoided, as the excessive amount of the iron in the diet is harmful [13]. It leads to some disease like hemochromatosis or siderosis. Thus, the aim of our research was to find a method of decreasing the absorbable iron content of fish silage.

The iron absorption is enhanced when soluble monomeric complexes are formed (in the presence of e.g. proteins, aminoacids, ascorbic acid, EDTA, citric acid etc.) [13].

The iron absorption is inhibited in the presence of carbonates, oxalates and phosphates. Probably they participate in macromolecular polymer formation, which are not absorbable from alimentary canal.

Thus, in our investigations, the sulfuric acid used in hydrolysate production has been replaced by phosphoric acid. Such a change should be effective in diminishing the iron absorption, as the iron phosphates are not soluble in water. In the first series of experiments, the hydrolyzates were produced only from fish, without sludge. The characteristics of hydrolizates produced from fish is presented in Tables 5 and 6.

Table 5. Characteristics of hydrolizates produced with using sulphuric acid or phosphoric acid

Process	Hydrolysis mixture	Protein [%]	Fat [%]	Water [%]	Ash [%]
1	Fish + 1 % H_2SO_4	14.1	6.6	76.1	2.7
2	Fish + 1 % H_3PO_4	14.0	6.2	77.0	3.0
3	Fish + 1 % H_2SO_4 + Fe*	14.2	6.4	76.8	3.1
4	Fish + 1 % H_3PO_4 + Fe*	14.2	6.2	75.2	3.8
5	Fish + 2 % H_3PO_4 + Fe*	13.9	6.5	74.1	4.2

* - samples enriched with iron - 200 mgFe/dm^3

Table 6. Iron content in fish hydrolyzates

Process	Hydrolysis mixture	Total iron [mgFe/kg]	Absorbable iron [mgFe/kg]
1	Fish + 1 % H_2SO_4	251.6	86.3
2	Fish + 1 % H_3PO_4	250.2	30.5
3	Fish + 1 % H_2SO_4 + Fe*	426.6	257.4
4	Fish + 1 % H_3PO_4 + Fe*	451.4	123.9
5	Fish + 2 % H_3PO_4 + Fe*	436.3	2.5

* - samples enriched with iron - 200 mgFe/dm^3

As it can be seen from Table 5, the hydrolyzate properties were similar in the case of both acids. The only difference was observed for absorbable iron content (Table 6). When phosphoric acid was used, the concentration of absorbable iron was much lower (30.5 mgFe/kg) than it was when sulphuric acid was applied (86.3 mgFe/kg). However, the decrease of absorbable iron was lower than it should be, considering the stochiometry. The amount of phosphoric acid used was about 20 times higher than that, required for complete iron precipitation in form of $FePO_4$. However, further addition of phosphoric acid (2 % process 5, Table 6) results in very effective decrease of absorbable iron content (2.5 mgFe/kg).

In the next series the hydrolyzates were produced from fish or fish viscera with the addition of 20 % of post-coagulation sludge. The characteristic of hydrolyzates is presented in Table 7.

Table 7. Characteristics of hydrolyzates produced witch using sulphuric acid or phosphoric acid

Process	Hydrolysis mixture	Protein [%]	Fat [%]	Water [%]	Ash [%]
1	Fish + 1 % H_2SO_4	14.1	6.2	76.4	2.6
2	Fish + 1 % H_3PO_4	14.3	6.3	76.9	2.9
3	Fish + sludge + 1 % H_2SO_4	12.2	6.3	77.4	3.1
4	Fish + sludge + 1 % H_3PO_4	10.9	6.3	78.0	4.2
5	Viscera + 1 % H_2SO_4	12.4	17.5	66.1	3.0
6	Viscera + 1 % H_3PO_4	12.6	17.8	67.7	3.3
7	Viscera + sludge + 1 % H_2SO_4	9.7	15.7	71.5	2.2
8	Viscera + sludge + 1 % H_3PO_4	9.5	14.9	72.4	2.9

As it can be seen, the sludge addition decreases protein content in silage by about 2 to 3 %. Other parameters have changed not much. The only difference was again observed for iron content (fig.2).

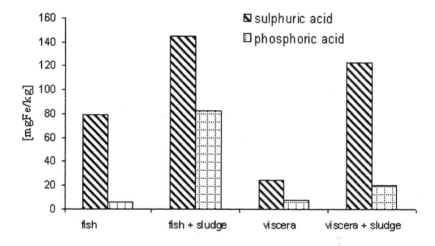

Figure 2. Concentration of absorbable iron in fish hydrolysates obtained with sulphuric or phosphoric acid

Figure 2 shows that the content of absorbable iron was higher if hydrolyzates were produced with the addition of post-coagulation sludge. Very effective reduction of the soluble iron concentration has been achieved when phosphoric acid was used instead of sulphuric acid.

3.   CONCLUSIONS

The results obtained in these investigations show that it is possible to recover nutritious substances (protein and fat) from fish processing wastewater by coagulation. Post-coagulation sludge can be used as a supplementary material for fish hydrolyzates production. This liquid feed is usually used for fur animal or young pigs feeding. Too high content of absorbable iron can be reduced by phosphoric acid use in hydrolysis process instead of sulphuric acid.

REFERENCES

1. Usydus, Z., Bykowski, P.J., Characteristics of wastewater in the fish processing industry, *Bulletin of the Sea Fisheries Institute*, 1 (143), 1998, 63-66.
2. Konicka-Wocial, E., Usydus, Z., Bykowski, P. Attempt to Using Thermal Spills from Fish Processing for Fodder, in *International Conference on Analysis and Utilization of Oily Wastes AUZO'96*, Gdańsk, 1996, 43-48.
3. Marti, C., Roeckel, M., Aspe, E., Kanda, H., Recovery of proteins from fishmeal wastewaters, *Process Biochem.*, 29, 1994, 39.
4. Usydus, Z., Bykowski, P.J., The utilization of waste and raw materials obtained in wastewater from the fish processing industry, *Bulletin of the Sea Fisheries Institute*, nr 1 (143), 1998, 17.
5. Usydus, Z., Bykowski, P.J., Treatment of wastewater from the fish processing industry factories, *Bulletin of the Sea Fisheries Institute*, 1 (146), 1999, 73.
6. Kim, Y. H., *Coagulants and Flocculants. Theory and Practice*, Tall Oaks Publ. Inc., Littleton, Co, 1995.
7. Zięcik, M., Podeszewski, Z., Kołakowski, E., Investigations on liquid feed production from fresh water fish and sea fish, Zeszyty Naukowe Akademii Rolniczej w Szczecinie, zeszyt *Agricultura*, 22, 1966, 217.
8. Raa, J., Gildberg, A., Fish silage, *CRS Reviews in Food Science and Nutrition*, 16, 1982, 383.
9. Gildberg, A., Recovery of proteinases and hydrolases from fish viscera, *Bioresource Technol.*, 39 (3), 1992, 271.
10. Raa, J., Gildberg, A., Autholisis and proteolytic activity of cod viscera, *J. of Food Technol.*, 11, 1976, 619.
11. Kołodziejska, I., Pik, A., Proteolytic acticity of the viscera of the Baltic herring (Clupea harengus), *Bulletin of the Sea Fisheries Institute*, 1 (134), 1995.
12. Mędrzycka, K., Szczesiul, J., Possibility of using oily waste from coagulation of fish processing waste for fodder, *International Conference on Analysis and Utilisation of Oily Wastes AUZO'96*, Gdańsk, 1996, Materiały Konferencyjne, 145-150.
13. Layrisse, M., Martinez-Torres, C., *Enhancers for Intestinal Absorption of Food Iron, in The Biochemistry and Physiology of Iron*, Elsevier North Holland, Inc, 1982.

SUSTAINABLE DEVELOPMENT IN INDUSTRY BY CLOSING WATER LOOPS: TECHNOLOGICAL ASPECTS AND EXPECTED FUTURE DEVELOPMENTS

Wim H. Rulkens
Wageningen University and Research Centre, Department Agrotechnology and Food Sciences, Sub-department of Environmental Technology, PO Box 8129, 6700 EV Wageningen, The Netherlands. Tel. +31 317 483339, Fax + 31 317 485883, E-mail: wim.rulkens@algemeen.mt.wag-ur.nl

Abstract: Closing industrial water loops can substantially improve the environmental sustainability of industrial production processes. In general a closed water loop consists of a large number of separate treatment steps arranged in a logical sequence. The selection of separate treatment steps that, together, comprise a closed loop water system, is complex. Various complete treatment scenarios can be developed and designed to satisfy the requirements set for process and transport water and treatment of wastewater. A technical and economic evaluation, in combination with an environmentally sustainability assessment, is necessary to determine the treatment system which is most appropriate. Examples of closed industrial water loops are already realised among others in the pulp and paper industry, surface plating industry, textile industry, food industry, and greenhouse horticulture.

It is expected that in the very near future, the application of closed water loops will show an intensive growth, strongly supported by the further development of separate treatment technologies such as: anaerobic treatment, membrane bioreactors, advanced biofilm processes, membrane separation processes, advanced precipitation processes for recovery of nutrients, selective separation processes for recovery of heavy metals, advanced oxidation processes, selective adsorption processes, advanced processes for demineralisation, and physical/chemical processes which can be applied at elevated temperature.

Key words: Sustainability, water treatment, closing loops, technology

1.   INTRODUCTION

During the last three decades an increasing awareness is observed into the environmental problems caused by human activities[22]. These environmental problems are a direct risk for human health and the human survival potential. To challenge these problems, human activities, in particular agricultural and industrial production processes, industrial products, and transport, but also the individual way of living have to be made more sustainable. Sustainability does not only focus on the environment. It also involves economic aspects and aspects dealing with implementation. Implementation includes in this respect education, acceptance, human involvement into the application, and incorporation into the human mind.

Industry was and is in fact still a major polluter. However, looking to the highly industrialised western countries we can observe a strong reduction in pollution in the various individual industries. This not only holds for existing industries, but also for new industries. Looking to paper mill industry, textile industry, surface plating industry, food industry, and metal industry, we can see a very substantial decrease in pollution by each of the individual industries. However, we live in a world wherein economic growth is a key factor to get welfare and employment, not only in the highly industrialised developed countries, but also in the developing countries. Especially in the latter countries we can expect a substantial economic growth. A strong additional economic growth can also be expected from the growing world population. It will therefore be clear that a sustainable development in the industrial sector is a key factor for the achievement of an environmentally sound world where people can live and work without unacceptable risks.

There are several approaches to improve current industrial production processes in more sustainable ones. However, the degree of sustainability can vary strongly, as is also the case with the costs. In general we can range these approaches in the following order of increasing sustainability.

a) Good house keeping. Characterised as a relatively cheap activity with very often a high environmental benefit.

b) End of pipe treatment, to abate emissions of pollutants to the environment. Characterised as a necessary activity to protect the environment for pollutants. Gives in general a minor contribution to the real sustainability. Costs are relatively high.

c) Process improvement focused on treatment of separate waste and waste water flows. The aim is to recover valuable products for reuse, to decrease the total amount of waste or wastewater and/or to reduce the energy need.

In general the costs can be substantial but there may also be substantial economic and, of course, environmental benefits.

d) Clean technology. Although not always very clearly defined, this approach fits most optimal with the issue of sustainability. It is an approach which can be applied to design completely new production processes in which both the raw materials, the production process and very often also the design of the product is considered in an integrated way, with sustainability as the leading factor. In that respect the product design has to satisfy criteria regarding minimal use of energy and materials and a possibility for complete recycling. In this way we can achieve optimal environmental sustainability. Due to an efficient use of proper raw materials, utilities (energy, water), and the absence or nearly absence of end of pipe technology, this approach very often leads to a very cost-effective way of production.

In many industrial production processes water plays a key role. Water may be necessary for several purposes. It can be used as:

- Cooling medium
- Medium for steam production
- Transport medium for materials (reactants, essential chemicals, wastes) and energy
- Medium wherein chemical or microbiological reactions can be executed
- Cleaning medium
- Drinking water and water for sanitation purposes.

As already mentioned many industrial production processes are using large amounts of water. Water that finally ends up as wastewater which has to be discharged into the sewerage system or, after treatment, onto surface water. This way of using water requires also a large amount of water supply. Industry is the second largest water user, after agriculture, with an amount of approximately 25% of the global demand. It is clear that this way is in general not sustainable and is often also expensive, especially in areas where we have to deal with a shortage of high quality surface water or ground water. This is the reason that there is a strong tendency to reuse water in an efficient way.[1, 2, 3, 4, 5, 6, 15] This reuse is combined with a minimisation of the water need, recovery of valuable products, production of water of different quality (depending on the use) and an optimal integration of the required process water quality and the treatment of the wastewater. This finally leads to closed water loops in which intake of water and discharge of wastewater have been minimised to a level that is technically and financially feasible, environmentally relevant and which can also be implemented. Also the loss of valuable products and energy and the high costs of treatment of large amounts of water in many of the current production processes strengthen the interest in the application of closed water loops.

The development of closed water loops in the industry requires new water management concepts and improved or innovative treatment technologies. During the last decade a lot of effort has been made into the development and application of new integral concepts. In some industrial branches closed water loops are already applied successfully. Up to now these approaches very often have been started and developed more or less on a management level. Less attention is paid to the technological and operational level. Closed water loop systems, in which also recovery of valuable products and saving of energy are essential factors, require very often approaches of wastewater treatment techniques which are different from the ongoing standard wastewater treatment techniques.

The development and implementation of a closed water loop and the choice of the optimal system is in fact a complex problem. In closed water loops, the composition of the wastewater streams may completely differ from the composition of the wastewater of an industry with conventional water supply and wastewater treatment systems. Besides the temperature of the wastewater in a closed loop system may differ from that of conventional systems. And finally, the removal efficiency of a pollutant can also differ strongly from the removal efficiency of the pollutants in case of a conventional wastewater treatment system.

As already mentioned, in some industrial sectors closed or almost closed water loops are already used, however in general the applications and experiences are still limited[1, 2, 3, 4, 5, 6, 15]. In any given situation, several distinct possibilities exist to close a water loop more or less completely. However, integration in the production process is always a requirement. Basically this is a very complex problem. The general question can be put how to develop such a closed water loop in an optimal way. This paper will discuss briefly what approach has to be followed regarding the treatment technologies which should or can be applied. The following issues are successively assessed:

– Characterisation of wastewater streams. This aspect is of an essential importance for identification, selection and designing separate treatment steps (treatment techniques) which, in combination, can lead to a complete wastewater treatment system that also can supply the various types of process water required.
– Overview of the most important current wastewater treatment technologies (separate treatment steps). Special attention is given to the specific area of application of the different treatment steps. The starting point in the

description of these separate treatment steps is the type of pollutants which may be removed. Not only is the type of pollutants important, but also the physical state of the pollutants. Pollutants can be present as soluble pollutants or as colloidal or suspended particles.
- Expected future developments in wastewater treatment technology, especially with respect to innovative technologies.
- Identification and selection of the separate wastewater treatment steps required for constituting a complete wastewater treatment and process water system.
- The logic sequence of the identified and selected separate treatment steps in the complete treatment scenario.
- Identification and selection of the complete wastewater treatment system that is optimal in economical, technical and environmental respects.
- Further incorporation of closed water loops in a broader environmental frame work, such as, for example in an eco-industrial area.

   The aim of this paper is to provide a basis and a tool for designing specific and concrete closed water loops for various industrial production processes wherein several types of process water are used, and wherein several types of wastewater are produced.

2.   CHARACTERISATION OF WASTEWATER STREAMS

   The entire wastewater treatment scenario necessary for a closed water loop strongly depends on the amount, composition and temperature of the various process water streams necessary for the production process, the amount, temperature and composition of the various wastewater flows, and the fluctuations in these wastewater flows.

   Regarding the pollutants which may be present in a wastewater stream, the following subdivision can be used, based on the applicability of the various types of treatment steps:
- Suspended particles (organic or inorganic)
- Colloidal particles (organic or inorganic)
- Soluble substances
- Non toxic organic pollutants
- Toxic organic pollutants
- Salts, acids, bases (KCl, $NaCl$, $CaCl_2$, HCl, $NaOH$, etc)
- Heavy metals in ionic state (Zn^{2+}, Cn^{2+}, Ni^{2+}, Cd^+, Cr^{3+}, Cr^{6+}, etc)
- N-compounds (nitrates, nitrites, ammonia, proteins)

- P-compounds (phosphates)
- S-compounds (sulphides, sulphates)
- Carbonates
- Hydroxides

In addition, the wastewater (and of course the process water) is also characterised by:

- pH
- Redox potential
- Temperature
- Alkalinity
- Presence of viruses and pathogenic micro-organisms.

Organic compounds, sulphur-, phosphorous- and nitrogen-compounds may be present as soluble, suspended or colloidal particles. This also holds for salts, bases and heavy metals. Pretreatment can change the physical state of pollutants in the wastewater. This gives a possibility for guiding the treatment process.

The entire wastewater treatment system focuses on removing the various types of pollutants with the highest possible efficiency from wastewaters, synchronizing this effort with the production of compatible process waters. Efficiency is not only considered in technical respect, but also deals with the costs of the purification process and with aspects of sustainability such as the use of chemicals, additives, and energy, the recovery of valuable compounds for reuse, the emission of volatile pollutants and the production of final wastes.

3.   OVERVIEW OF WASTEWATER TREATMENT STEPS

Purification of wastewater can occur according to two basic processes:

- Removal of pollutants by means of physical processes
- Chemical or biological conversion of pollutants to compounds which are less polluting than the original ones, and completely non-polluting if possible.

Removal of pollutants from wastewater by physical processes results in the concentration of these pollutants in a small volume. The molecular composition of the pollutants is not changed in such a process. Dependent on the type and concentration of the pollutants the concentrate may have subsequently to be treated. Several options are available for that purpose:

- Destruction of the contamination (if the pollutants are of an organic nature)
- Reuse as a valuable compound
- Storage in a controlled disposal site.
  Two types of processes are possible for physical separation:
- Processes based on phase separation. These processes are appropriate if pollutants are present as colloidal or suspended particles.
- Processes based on molecular separation. These processes are appropriate if pollutants are present in soluble state. A molecular separation process is consequently characterised by the use of a second phase. Separation can be achieved by contacting the wastewater with a gas, solid or non-aqueous liquid phase (not mixable with water) which has a higher affinity to the pollutants than the water phase. The pollution will then be concentrated into this phase. This second phase can also be created by heating the wastewater, so that part of the water phase evaporates, or by cooling the wastewater, which results in the formation of ice crystals. With these processes it may be possible, depending on the type of pollutants present in the wastewater, to remove the water more or less completely. By changing the temperature or adding chemicals it is also possible to transform soluble pollutants into insoluble compounds, creating a separate phase which can be separated form the wastewater.

Conversion processes aim to transfer a pollutant into a compound, which is less, or non-polluting to human beings and ecosystems. Two types of conversion processes are distinguished:

- Chemical conversion, that can occur by adding appropriate chemicals to the waste water, which react with the pollutants
- Microbiological conversion processes, typically using microorganisms, which degrade the pollutants to compounds which are less polluting.
  There are two types of microbiological conversion processes: anaerobic processes and aerobic processes.

All types of separate wastewater treatment steps are based on one or more of the above-mentioned basic process steps. In practise a large number of treatment steps are applied, some on a large scale, others on a smaller scale.[1, 2, 3, 4, 5, 6, 7] In the following an overview is given of the most important separate treatment steps. The starting point in this overview is the type of pollutant and the physical state of the pollutant. The overview is aimed as a brief indication of separate treatment steps which are potentially applicable to remove the specific type of pollutant. More detailed information about principles, possibilities of application, removal efficiencies, bottlenecks and costs are present in the literature, including also handbooks on wastewater treatment.

3.1   Removal of suspended particles

- Settling/sedimentation (followed by separation and additional dewatering of the sludge)
- Flotation (floating of the pollutants by attachment of small air bubbles to the suspended polluted particles)
- Sieving/microsieving (removal of relatively large particles)
- Filtration (surface filtration and deep bed filtration)
- Centrifuging (separation using centrifugal forces)
- Hydrocycloning (separation using centrifugal forces)
- Magnetic separation (separation of ferro- or paramagnetic particles using magnetic forces)
- Microfiltration (filtration across a semi-permeable membrane at low pressure drop).

It has to be noted that several of the mentioned techniques, for example flotation, require the use of coagulation/flocculation agents to promote conglomeration and agglomeration of small particles to larger ones. Several modifications of each technique exist. Some of the techniques are only applicable if the concentration of suspended particles is not too high. In all applications sludge is produced which needs further treatment, such as dewatering. The quality of the treated water with respect to residual amounts of suspended particles is strongly influenced by the type of wastewater, the type of technique, and the applied process conditions.

3.2   Removal of colloidal particles

- Basically the same types of techniques are applicable as were mentioned for the removal of suspended particles, provided there is also a pretreatment step in which the colloidal particles have been destabilised and have agglomerated or conglomerated to larger particles. This destabilisation process can occur by adding coagulating and/or flocculating agents.
- In addition to the above mentioned techniques, the following techniques can be applied:
- Chemical oxidation (in the case of organic compounds)
- Adsorption onto biomass, eventually followed by an aerobic or anaerobic biodegradation process.

3.3 Removal of non-halogenated soluble organic compounds

- Anaerobic biodegradation (production of biogas)
- Aerobic biodegradation (conversion into CO2, water, and biomass)
- Adsorption to activated carbon or another type of adsorption agent
- Chemical oxidation using ozone, hydrogen peroxide, potassium permanganate or sodium hypochlorite (in the case of relatively low concentrations)
- Stripping with steam, air or under vacuum (followed by a further treatment of the stripping gas). Applicable to volatile pollutants
- Separation and concentration by means of hyperfiltration (reverse osmosis).

Each type of technique can be applied in several modifications. Microbiological degradation processes offer the possibility to purify wastewater streams which contain a wide range of organic pollutants. In contrast, physical/chemical techniques are much more specific. The quality of the purified water regarding residual amounts of soluble organic pollutants, and colloidal and suspended particle pollutants, strongly depends on the type of treatment process and the applied treatment conditions. Very often a combination of different types of treatment techniques is necessary to satisfy the effluent quality required (for example a combination of anaerobic and aerobic treatment).

3.4 Removal of soluble toxic halogenated organic compounds

- Anaerobic pretreatment in combination with aerobic posttreatment
- Electrochemical dehalogenation (applicable to pentachlorophenol, trichlorobenzene, hexachlorobenzene, etc.)
- Adsorption to activated carbon (or another type of adsorbent)
- Stripping of volatile pollutants (by air, steam or under vacuum) followed by a treatment of the stripping gas
- Oxidation using ozone, potassium permanganate, hydrogen peroxide or sodium hypochlorite
- Separation by reverse osmosis.

It has to be noted that the various processes mentioned are especially aimed at removing or destroying halogenated organic pollutants. If, in addition to halogenated pollutants, large amounts of non-halogenated organic pollutants are also present, the overall effectiveness of these processes is reduced.

3.5   Removal of soluble minerals

- Precipitation, followed by the removal and further treatment of the concentrate
- Crystallisation (several modifications)
- Ion exchange. In general very selective. Regeneration of the ion exchange media is necessary, as is treatment of the concentrate
- Electrodialysis (membrane process based on ion-selective membranes). In this process a concentrate is produced that needs further treatment
- Hyperfiltration
- Evaporation (selective removal of water)
- Freeze concentration (selective removal of water).

The quality of the purified water and the composition of the concentrate strongly depend on the type of techniques and the applied process conditions. The presence of suspended and colloidal particles or dissolved organic compounds can have a negative effect on the treatment.

3.6   Removal of heavy metal ions

- Essentially all the same treatment steps can be used as were already mentioned for the removal of mineral salts
- In addition to these treatment steps, or as a specific modification of these treatment steps, various other treatment techniques are available such as:
- Electrolysis, resulting in a recovery of the heavy metal ions as pure metal
- Cementation
- Adsorption to activated carbon
- Selective precipitation of metal ions using hydrogen- or sodium sulphides, hydroxides or carbonates
- Use of ultrafiltration membranes after complexation of the heavy metal ions.

Also these processes can be applied in different modifications. The complete removal and recovery of heavy metals often requires the application of a combination of several treatment steps.

3.7   Removal of dissolved nitrogen compounds

- Microbiological conversion in molecular nitrogen by means of nitrification and denitrification, often in combination with the removal of organic pollutants
- Precipitation of ammonia to struvite
- Stripping of ammonia by air, steam or under vacuum, followed by treatment of the ammonia containing gas phase
- Ion exchange for removal of ammonia, nitrite and nitrate
- Electrodialysis for treating ionic nitrogen containing compounds
- Hyperfiltration (reverse osmosis)
- Anaerobic treatment. In anaerobic processes the organic nitrogen compounds are converted to ammonia.

3.8   Removal of soluble phosphate

- Precipitation into iron-, calcium- or aluminium phosphate
- Crystallisation
- Microbiological concentration in biomass
- Hyperfiltration
- Magnetic separation (after precipitation as iron phosphate or after adsorption to magnetite)
- Adsorption.

3.9   Removal of dissolved sulphur compounds (sulphates, sulphides, organic sulphur compounds)

- Precipitation as calcium sulphate or iron sulphide
- Crystallisation
- Hyperfiltration
- Ion exchange
- Stripping (hydrogen sulphides) followed by treatment of the stripping gas
- Anaerobic treatment. Anaerobic processes can convert organic sulphur compounds and sulphate into sulphides
- Partial oxidation of hydrogen sulphide into insoluble elemental sulphur
- Chemical oxidation of sulphides into sulphates.

3.10   Decolourization of wastewater

- Use of oxidation agents such as ozone, hydrogen peroxide, very often in combination with UV and/or a catalyst
- Adsorption
- Anaerobic and aerobic biological degradation
- Membrane filtration (reverse osmosis, nanofiltration, ultrafiltration, microfiltration)
- Flotation
- Electrochemical treatment.

3.11   Disinfection of the wastewater

- Use of oxidation agents such as ozone, hydrogen peroxide, sodium hypochlorite
- Use of UV
- Heating
- Gamma radiation
- Membrane filtration (ultrafiltration, nanofiltration).

This brief overview of separate treatment techniques is not complete, however it can be used for a first inventory and identification of treatment steps which may be considered as part of a complete closed loop water system. It has to be taken in mind that the above-mentioned overview of separate treatment techniques is primarily based on one type of pollutant and one physical state of that pollutant. It will be clear that very often the same treatment step can be applied to remove different types of pollutants. It is also evident that a large percentage of the separate treatment steps mentioned will result in a concentrate containing the pollutants. This concentrate has to be treated subsequently.

Costs of treatment techniques vary strongly. The costs not only depend on the type of treatment process but also on the concentrations of the pollutants to be removed, and the required removal efficiency of the pollutants. Most treatment steps have been developed for the treatment of wastewater at a temperature above zero degree Celsius as lower limit and somewhat above ambient temperature as upper limit. An exception is anaerobic treatment, which is often also applied at temperatures of about 60°C. In general little or no experience is available with the purification of wastewater at temperatures above 50° C. Experimental research is necessary to assess the feasibility of a treatment step in that temperature range.

4.   EXPECTED FUTURE DEVELOPMENTS IN WASTEWATER TREATMENT TECHNOLOGY

The survey of separate wastewater treatment technologies as given in the previous paragraph clearly shows the large number of treatment steps and the variation in treatment processes available for solving wastewater problems. It is also obvious that for each wastewater problem, characterised by the amount and composition of the wastewater and the required removal efficiency for the various components, many treatment steps and treatment chains (combination of treatment steps) are available. Which types of the current treatment processes are most optimal for solving an existing wastewater problem depends on various factors such as required efficiency, costs, energy need, waste production, etc. There is of course a strong and continuous stress for further improvement of existing processes and the development of new, cheaper, more effective, and more sustainable processes. The development of new wastewater treatment technologies and new integrated concepts of wastewater treatment chains is in fact a dynamic process, partly governed by the needs of the market.

It may be expected that due to the shift to cleaner industrial production processes the amount and composition of industrial wastewater will change. As most important changes can be mentioned:

- Use of less amounts of water and therefore also a lower production of wastewater. In some cases this will result in higher concentration of the pollutants.
- Production of wastewater with less variety of types of pollutants.
- Wastewater of a higher temperature
- The absence of real toxic pollutants such as cadmium, mercury, PCB's etc. due to the policy to ban the use of these components in industrial production processes.

Also the aim of the wastewater treatment process will change strongly. Preference will be given to treatment technologies which have not only the potential to purify water but which have also the potential to produce valuable products from pollutants (such as biogas) or to recover valuable components from the wastewater, such as nutrients (phosphate and ammonia), heavy metals or specific minerals. Also the specific process conditions, such as use of chemicals and energy, and the wastes which are produced in a wastewater treatment step are more and more considered from an environmental point of view. Finally it can be expected that the compactness of the treatment process, related tot the required residence time of the wastewater in the system, is becoming more and more important. The future development of new (separate) wastewater treatment steps will follow this tendency.

As already mentioned a wastewater treatment step is always based on one or more of the four types of fundamental processes: phase contact and phase separation processes, molecular separation processes, chemical conversion processes, and microbiological conversion processes. The development of new wastewater treatment steps occurs in fact according to these fundamental treatment steps. That means that the future development in the field of wastewater treatment technology is in fact also strongly influenced by other fields of environmental, microbiological, chemical and technological sciences. Of course, also the practical experience with current wastewater treatment technologies contributes to the improvement of existing technologies and the development of new technologies. More specific the expected future development will strongly be based on the developments we can expect in the following scientific fields:

- Knowledge of microbiological conversion processes which takes place in aquatic and terrestic ecosystems. Here we can often identify and isolate the micro-organisms which are adapted to the pollutants of concern and which are responsible for the biodegradation. In fact also the current biological treatment processes are almost completely based on micro-biological degradation processes occurring in nature. However, also from the industry dealing with the microbiological production of chemicals, pharmaceuticals, pesticides, etc. we can learn a lot.

- Knowledge about chemical conversion processes which take place in aquatic and terrestic systems. Also the knowledge and experience of the chemical industry may be of interest in the development of new chemical treatment processes.

- Technology development in the process industry, the chemical industry and in the industry dealing with biotechnology

Regarding microbiological processes for treatment wastewaters we can expect the following innovative developments[12, 13, 14]:

a) Further development of anaerobic processes. This process is cheap and very energy efficient. It can be applied in a broad range of temperatures. It is also very effective in the conversion of sulphur components, present in water or air, into elemental sulphur, which may be an interesting product for reuse. Anaerobic treatment is especially of interest when dealing with high concentrations of organic pollutants, which have no direct value for reuse.

b) Membrane bioreactors. This process combines the advantage of a microbiological conversion with the advantage of a microfiltration or

ultra-filtration membrane to separate and concentrate the micro-organisms. The technology has the advantage of compactness and the production of an effluent free from colloidal and suspended particles.
c) Advanced biofilm processes. In these processes the micro-organisms are immobilised on a fixed or suspended carrier material. This system has also the advantage of compactness.

With respect to physical/chemical treatment processes for industrial wastewater we can expect the following innovative developments[11, 16, 17, 19, 20, 21].

4.1 Membrane separation processes

By far membrane processes are the most promising innovative processes for treatment of wastewater, aimed at reuse and the production of a high quality effluent. At present it is a technology that is already world-wide in operation on a large scale for the production of drinking water and the treatment of industrial wastewater. In general the standard membrane processes can be classified in four groups: microfiltration, ultrafiltration, nanofiltration and hyperfiltration or reverse osmosis. The first two processes are characterised by the property that water and dissolved low molecular pollutants can pass the membrane while colloidal and suspended particles are retained by the membrane. The permeability for dissolved high molecular weight pollutants depends on the size and pore structure of the membranes. Nanofiltration and hyperfiltration membranes are permeable for water but can also retain dissolved pollutants of lower molecular weight. The retention strongly depends on the pore size and membrane structure and can be more than 99%. Also viruses and other pathogens can be retained by these membranes.

A bottleneck in all membrane processes, applied in practice, is fouling and scaling of the membranes. These processes cause a decrease in water flux through the membrane and a decrease in retention. Much attention is paid, especially in case of nanofiltration and hyperfiltration, to prevent fouling of the membrane by an intensive pretreatment and the regular removal of fouling and scaling layers by means of mechanical, physical or chemical treatment.

There is no doubt that membrane processes are becoming one of the most powerful and important processes in wastewater treatment and water reclamation. The main reasons for this expectation are the following:
- During the last ten to fifteen years a substantial operational improvement of fluxes through the membranes and retention of the membranes has

been observed, accompanied by a continuously decrease in membrane costs. This tendency will continue in future.

- Regarding the retention of pollutants membranes can be manufactured more or less tailor made, thus increasing potential applications
- Membrane processes can not only produce a water quality that satisfies set standards but can simultaneously concentrate the pollutants in a small volume. This strongly promotes the possibility for recovery of pollutants such as ammonia and phosphate for reuse. Also biological conversion of dissolved organic pollutants becomes easier.
- Membrane processes, such as hyperfiltration, can also reduce the health risks caused by the presence of pollutants such as pesticides, heavy metals, endocrine disrupters, pathogens and viruses. This is especially important in case where human beings can get in contact with reclaimed water.

4.2   Precipitation processes.

To recover valuable inorganic products from concentrated or diluted wastewater streams very often precipitation (and crystallisation) processes are applied. The aim of these precipitation processes is to produce a granular solid material that can be reused or processed elsewhere. Precipitation almost always requires the addition of chemicals or the decrease (sometimes increase) in temperature or the removal of water. The process operates most adequately if we have to deal with a concentrated wastewater. Therefore very often a pretreatment of the wastewater is required. This pretreatment may include an ad(b)sorption of the valuable component, for example by an ion exchanger, or a removal of water by means of hyperfiltration, freeze crystallisation or evaporation. Examples of interesting precipitation processes for recovery of valuable products from wastewater are the precipitation of phosphate and ammonia from wastewater. Recovery of phosphate is relevant because of the future expected shortage of phosphorus containing raw materials. Recovery of ammonia instead of biological conversion to nitrogen is of interest because the production of ammonia requires an energy source and the biological conversion of ammonia to nitrogen also needs an energy source. To recover dissolved phosphate from a wastewater several precipitation processes are available. One process is the precipitation of phosphate as calcium phosphate by adding lime to the wastewater. Several modifications of this process exist. One modification is the DHV Crystalactor. In this process a fluidised bed reactor is used in which calcium crystallises on the bed particles. The process produces pellets which for a large part consist of calcium phosphate and which can be used as a source for the production of phosphorous. Another process is the precipitation of phosphate as struvite. Struvite is a relatively insoluble precipitate with an equimolar ratio of magnesium, ammonium and phosphate ions (Mg NH$_4$ PO$_4$.6 H$_2$O). Precipitation of phosphate as struvite from wastewater containing phosphate and ammonia requires the addition of magnesium oxide or magnesium hydroxide. The solubility of struvite depends on the pH. The crystallisation process is also strongly influenced by the presence of other inorganic and organic compounds in the wastewater. Basically struvite can be used as fertiliser. Struvite precipitation is only suitable for wastewater streams which contain relatively high concentrations of phosphate. Wastewater which contains low concentrations of phosphate needs a concentration step as pretreatment. To that aim ion exchange processes or membrane processes such as hyperfiltration and electrodialysis can be used. Precipitation of ammonia from the wastewater may occur also according to the struvite route. A prerequisite, however, is that the ammonia concentration is high enough. A concentration step, for example reverse osmosis, is often necessary as pretreatment step.

4.3   Selective precipitation of heavy metals.

Industrial activities in the past have led to sever heavy metal contamination of ground water. Metal pollution still continues with the discharge of industrial wastewaters from metal processing, metal finishing and plating industries and acidic mine drainage waters. There are numerous treatment technologies for the removal of heavy metals from wastewater including precipitation, ion exchange, solvent extraction, complexation, adsorption, filtration and membrane processes. Most of these treatment technologies produce concentrated sludges that have to be stored in controlled hazardous waste disposal sites. A more sustainable solution for heavy metals containing wastewaters could be obtained if treatment processes focus more on recovery and reuse of the metals. Reuse of metals can in general only become economically when metals are removed selectively and mono metals or metal sludges are produced. Besides, there is also an increasing need for treatment systems able to achieve very low effluent concentrations. A precipitation method which can satisfy these two conditions is precipitation by sulphide. Using a sulphide ion-selective electrode a selective removal of heavy metals as heavy metals sulphides is possible. Use of a microfiltration or ultrafiltration can improve the selectivity of the separation process and can also concentrate the metal sulphide precipitate. The metal sulphides can be processed in existing smelters for heavy metals.

4.4   Advanced oxidation processes.

During the last two decades an increasing interest in the application of Advanced Oxidation Processes (AOPs) is observed. These processes are applied for the oxidation of toxic organic pollutants present in wastewater or in surface water that is used for the production of drinking water. The process of oxidation is mainly based on the oxidative destruction by radicals of which the hydroxyl radical (OH$^{\bullet}$) is the most powerful one. Most important AOPs are:

- Ozone
- Hydrogen peroxide
- Ozone in combination with hydrogen peroxide
- UV in combination with hydrogen peroxide
- UV in combination with hydrogen peroxide and ozone
- UV and TiO$_2$ as catalyst for the oxidation process

The advanced oxidation processes offer the possibility for destruction of recalcitrant biodegradable and toxic organic pollutants. The field of applications regarding wastewater treatment varies. AOPs can be applied as:

- Polishing step for removal of rest amounts of toxic and non-toxic organic pollutants
- Pretreatment process, focused on the selective oxidation of toxic pollutants which may disturb the biodegradation of the non toxic organic pollutants
- Treatment process for partial oxidation of recalcitrantly biodegradable organic pollutants. Aim is to make this recalcitrantly biodegradable products biodegradable in conventional aerobic and anaerobic wastewater treatment systems

AOPs are less appropriate for the complete treatment of wastewater streams containing high concentration of organic pollutants. The main reason is that the energy costs and costs of chemicals such as ozone and hydrogen peroxide are relatively high. In case of UV also the equipment costs may be substantially. To treat these concentrated waste streams the application of AOPs has to be focused on the selective oxidation of specific toxic pollutants or on the partial oxidation of pollutants.

4.5   Use of physical/chemical processes for treatment of hot waste wastewater.

In many production processes the process water is used at a relatively high temperature. Also the wastewater which comes free may have an elevated temperature. To treat this wastewater with standard physical/chemical treatment processes at present this wastewater is cooled down to ambient temperatures. In case the treated water can be reused as process water it would be more efficient if the wastewater could be treated at higher temperatures, for example at 60 to 80 °C. However experience with the use of physical/chemical processes in this temperature range is lacking. It can be expected that in future more attention will be paid to this knowledge gap.

4.6   Other physical/chemical processes.

Besides the processes mentioned at a. to e. there are some other physical/chemical processes which may be of interest in future to be used in closed water loops. These processes are
- Selective ad(b)sorption of compounds from wastewater, focused on recovery
- Advanced demineralisation processes in order to tackle the problem of the water soluble minerals.

5.   PREDESIGN OF AN INDUSTRIAL CLOSED WATER LOOP

As a first approach it is assumed that an industrial closed water loop focuses on the production of one type of process water of a constant water quality and temperature. After this process water has been polluted and has become a wastewater, it will be treated to the required quality and will be used again as process water. In fact this is the simplest form of a closed loop water system. We will discuss this situation first. The first step in the design of such a simple closed loop water system is to consider the type of separate treatment steps and the sequence of these treatment steps which have to be applied on a wastewater stream that contains large amounts of easily biodegradable soluble pollutants in addition to non-biodegradable soluble organic and inorganic pollutants and suspended and colloidal particulate pollutants. The general sequence of these treatment steps is shown schematically in Figure 1.

The first step generally consists of a pretreatment focused on the removal and concentration of suspended and/or colloidal particles. Depending on the type and concentration of the particles to be removed the following treatment steps (among others) may be applied as pretreatment:

- Sedimentation
- Flotation
- Sieving
- Filtration
- Hydrocycloning
- Centrifuging.

Often the addition of specific chemicals is necessary to obtain a practical, effective process. The intensity of the pretreatment strongly depends on the pollution level that is acceptable for the anaerobic and/or aerobic treatment steps, following the particle removal step. Easily biodegradable colloidal pollutants do not cause a problem. However inorganic particles can cause serious problems in biological treatment steps and need to be removed first.

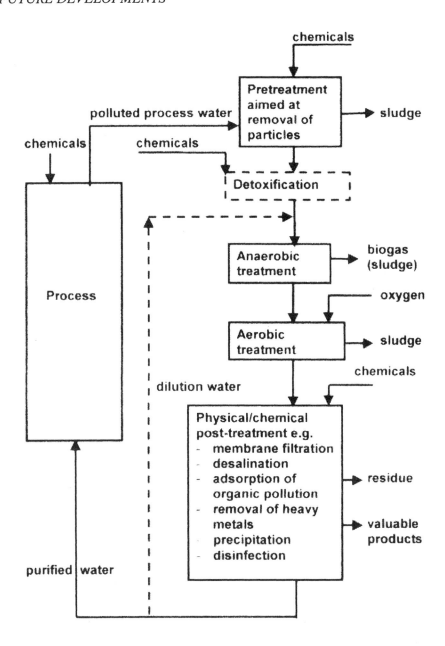

Figure 1. Industrial closed water loop (simplified process)

For the assumed type of wastewater this pretreatment step is very often followed by an anaerobic treatment step. This is especially the case if the wastewater contains a high concentration of easily biodegradable organic compounds. Production of biogas in the anaerobic treatment step can save energy. In the anaerobic treatment step organic nitrogen containing pollutants are converted into ammonia.

Dependent on process conditions sulphates may be biodegraded into sulphides. If necessary, the removal of ammonia and hydrogen sulphide can be achieved, after adjustment of the pH, by stripping. Sulphides can also be removed by precipitation with iron salts.

In the aerobic treatment step, which in general follows the anaerobic treatment step and the specific treatment processes for removal of ammonia and hydrogen sulphide, residual amounts of organic pollutants are removed. A proper sequence of process steps and proper process conditions in the aerobic treatment step can also lead to the removal of nitrogen and phosphorus containing pollutants

The physical/chemical treatment steps, as final treatment steps, are focused to remove small amounts of organic pollutants, soluble salts, heavy metals, pathogens and microorganisms. The physical/chemical treatment results in residues that need further treatment. The purified wastewater can be reused as process water. In case toxic compounds are present in the wastewater, it may by necessary to include a detoxification step before the biological treatment, in this case before the anaerobic treatment step. Various toxic compounds can be considered, such as heavy metal ions or toxic organic compounds, such as pesticides. Heavy metals can be removed by precipitation with hydroxides or sulphides, assuming that the concentration of colloidal and/or suspended particles is low. Another possibility for removal of heavy metals is electrolysis. In this way a possible disturbance of the anaerobic treatment step is prevented, as well as the production of heavy metal containing surplus sludge. Detoxification of wastewater containing toxic organic pollutants is possible by applying advanced oxidation processes using ozone, hydrogen peroxide, UV, or combinations of these treatment steps. A reduction in toxicity of the wastewater can also be realised by dilution of the wastewater with purified wastewater.

The types of physical/chemical (post)treatment steps which have to be applied strongly depend on the required quality of the process water (type and concentration of pollutants). Very often the following sequence is used:

- Removal of residual amounts of organic pollution. To that aim activated carbon or oxidation with ozone or hydrogen peroxide can be used

- Precipitation of heavy metals with hydroxides, carbonates or sulphides. Selective precipitation of the different types of heavy metals is also possible
- Desalination using ion-exchange, electrodialysis, or reverse osmosis
- Disinfection with ozone, hydrogen peroxide or hypochlorite.

Very often the use of process water and/or the treatment of wastewater requires the use of inorganic or organic additives. Sometimes these additives have to be considered later on as pollutants, sometimes as compounds of which the concentration in the process water should not be too high. It is therefor not allowed that the concentration of these additives exceeds a maximal admissible value. This means that in case of a continuous closed loop system, the physical/chemical treatment step has to remove an amount of these compounds corresponding to the amount of these compounds added elsewhere in the closed loop water system, minus the amount consumed in the production process. Often the consequence is that only relatively low removal efficiency in the treatment step is necessary.

In the aforementioned closed loop water system the polluted process water is purified completely in one wastewater treatment scenario. Only one type of process water is produced for reuse. In practise several types of process water are needed, and also several types of polluted wastewater streams are produced. This is schematically shown in Figure 2. In that case it can be advantageous in technical, economic and environmental respects to include several closed loop water systems within one production line.

In all closed loop water systems final residues such as sludges and concentrates are produced. These residual streams need further treatment. An appropriate selection of treatment steps, sequence of treatment steps, and process conditions to apply in each treatment step can result in recovery of valuable products for reuse. However additional and sometimes modified process steps may also result in an increase of treatment costs. These extra treatment costs have to be weighed against the benefits of the valuable products.

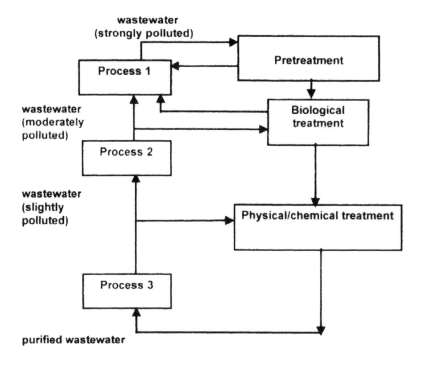

Figure 2. Closed water loop (several types of process water and wastewater)

A complete closed water loop is not always the most optimal way to deal with process water and wastewater in technical, economical, and environmental respects. It may be useful to discharge part of the wastewater after a partial clean up. For example, discharge may be appropriate if salt has to be removed and the discharge of purified wastewater to surface water allows the presence of a certain amount of salts, or if small amounts of organic pollutants and nutrients can be removed by natural degradation in the environment. It has to be emphasised again that this approach can only be followed if the set standards for discharge to surface water are satisfied. Partial discharge of incompletely purified wastewater into surface water requires an external supply of water. This water can be of drinking water quality. The general advantage of such an approach of the water loop is that the purification efficiency of the several wastewater treatment steps can be less than in case of a completely closed water loop, resulting also in lower treatment costs. It is also possible that compounds, which are added to the water, and/or which, arise in the wastewater during the production process, finally end up in the product. This is, for example, very often the case in board paper production processes. In such situations the requirements of the purification process will be less strict.

In a number of industrial production processes water is used at a relatively high temperature of 60 to 70°C. A first approach to purify the wastewater is to consider the application of treatment steps in that temperature range. However, except for anaerobic treatment and, also to a limited extent for aerobic treatment, there is little or no experience with the application of wastewater treatment steps at such high temperatures. It is of course possible to cool the wastewater to ambient temperatures, then treat this wastewater with existing conventional wastewater treatment steps, and reheating the purified waste water to the required temperature, so it can be reused as process water. However, this procedure means an extra capital investment for heat exchange equipment and the loss of energy. These disadvantages have to be compared with the possible complications and disadvantages of a purification process at a higher temperature.

6.   OPTIMAL SELECTION OF A CLOSED LOOP WATER SYSTEM

In the foregoing paragraphs it was shown that an industrial closed loop water system can be realised in several ways. Alternatives which can be selected are:
- A completely or a partially closed loop water system
- Types of waste water treatment steps which have to be applied
- Sequence and combination of these separate treatment steps
- Temperature at which the various treatment steps will be applied
- Application of one or more closed water loop systems.

The crucial question is always what system is most optimal. To answer this question, the following aspects have to be considered:
- Technical feasibility
- Environmental sustainability
- Economical feasibility.

6.1   Technical feasibility

From a technical point of view the proposed closed water loop has to be technically feasible. Risks of failure have to be as low as acceptable. In that context the following factors are of a high importance:
- Treatment steps that are included in the closed loop system
- Sequence of the various treatment steps
- Stage of development of the treatment steps
- Experience obtained with the various treatment steps
- Reliability of the complete system
- Sensitivity to disturbances and fluctuations in flow and composition
- Possibility to include the closed loop system in the production process
- Required expertise to operate the installation.

In general, it can be expected that several types of closed loop water systems will be technically feasible for any particular industrial facility.

6.2   Environmental sustainability

Closing a water loop is primarily focused on the achievement of a more environmentally sustainable production process. The treatment scenarios that have been identified as technically feasible should be further evaluated on criteria of environmental sustainability. Within the context of closing industrial water loops the following environmental effects have to be considered at a minimum:
- Savings in water use
- Savings in energy use/recovery of energy
- Savings in the use of non-renewable raw materials
- Prevention of the emission of pollutants to the environment
- Prevention of the production of final wastes.

A broader and more detailed evaluation can be done by performing a Life Cycle Analysis (LCA). The central idea of a LCA is that the environmental effects during the entire life cycle of a process are quantified. These environmental effects are caused by the use of fossil fuels for heating and production of electricity, the use of non-renewable raw materials for the production of materials and chemicals, and the emissions of pollutants to air, water and soil. These environmental effects can be subdivided further in various levels of detail. The five major effects mentioned are derived from the more general effects considered in the framework of the LCA. Based on the environmental sustainability of each of the complete treatment scenarios considered as technically feasible, a ranking according environmental sustainability can be obtained. In this way a selection of treatment scenarios which are feasible, both in technical respects and environmental respects, can be made.

6.3   Economical feasibility

The scenarios that have been selected as most optimal in terms of technical and environmental criteria finally have to be assessed with respect to financial consequences. Capital costs and operating costs have to be compared with the costs that are considered financially feasible.

In general it is not possible to develop a complete treatment scenario that is optimal technically, financially and also with respect to environmental sustainability. A solution has to be found which is economically and technically feasible and also satisfies the criteria of environmental sustainability to the greatest possible extent. An additional criterion for the selection of a complete closed loop water system can be obtained if the net environmental benefits are compared with the extra costs of the closed loop system. In this way the extra environmental benefits per unit investment costs or operating costs can be calculated. In the final selection of a closed loop water system aspects dealing with acceptability and public environmental awareness also play an important role.

7.   DISCUSSION

Closing industrial water loops can substantially improve the environmental sustainability of industrial production processes. In general a closed water loop consists of a large number of separate treatment steps arranged in a logical sequence. Very often a closed water loop includes several sub-loops focused on one or a few specific steps in the production process. Given the large number of separate treatment steps that can be applied in a complete water purification process, and the technical, economical and environmental aspects of each step, it has to be concluded that there is no single unique treatment scenario that is optimal in all respects. In general several alternatives satisfy these three criteria to various degrees. The selection of the alternative that fits most optimally to the production process depends strongly on the set boundary conditions and starting points. Regarding environmental sustainability, factors such as water savings, saving of chemicals, recovery of valuable components and energy from the wastewaters, savings on energy use, prevention of emissions and pollutants, and prevention of waste products have to be considered. It is impossible to design a closed water loop system that satisfies all mentioned environmental aspects in an optimal way.

The starting point in the development and designing of a closed water loop system is an inventory of the amounts and the quality of the process and transport water flows which are needed for the various steps in the production process. Each production step where process or transport water is involved causes a certain amount of wastewater. The pollution of this water is strongly dependent on the process step. The selection of separate treatment steps which, together, comprise a closed loop water system is complex. As already mentioned, various complete treatment scenarios can be developed and designed to satisfy the requirements set for process and transport water and treatment of wastewater. A technical and economic evaluation, in combination with environmental sustainability assessment, is necessary to determine the treatment system which is most appropriate.

During the last three decades the increasing awareness and concern in industry regarding environmental pollution has resulted in a tremendous reduction in pollution. It started with the application of end of pipe technologies and shifted gradually to a more general approach focused on environmentally sustainable production process. In some industrial branches but also in agricultural production systems this has already resulted in increased water reuse and the closing of water loops combined with recovery of valuable components. Examples of this development are

- Pulp and paper industry[9, 10]
- Surface plating industry
- Textile industry[8,18]
- Food industry
- Greenhouse horticulture

Closing the water loop is, in general, a first important step to achieve more sustainable production processes. A further improvement of the environmental sustainability of the production process can be obtained by an environmental assessment of all material and energy flows in the production process. This gives a more complete picture of possible ways to save on materials and energy use, and also how to minimise the amount of final wastes from the production process.

Regarding costs, it can be expected that closing water loops will often result in lower costs of the production process and in savings of materials and energy and in lower discharge costs for wastewater. However, it is unavoidable that residual waste streams are produced which have to be disposed in an environmentally acceptable way, or which need additional treatment. This may result in extra costs. These additional costs have to be considered in relation to the extra environmental benefits due to the use of a closed loop water system. Dependent on the costs and environmental benefits, it may be more attractive in certain cases to close the water loop only partially, and to discharge a small part of the wastewater after a more or less intensive treatment into the sewerage system or into surface water. Of course this has to be done according to the legal guidelines for discharge of treated wastewater to the sewerage system or to surface water.

A further improvement in the sustainability of an industrial production process may be achieved by closing the water loops for a number of adjacent industrial production processes. This may be achieved in a so-called eco-industrial area where the primary aim is to arrange industrial production processes in such a way that water, wastes, materials, and energy can be exchanged between the various production processes in an environmentally sustainable and cost effective manner. A shared process water production plant and wastewater treatment plant is then a crucial step. It will, however, be clear that such an approach requires a thorough and detailed study which incorporates not only technical and economical aspects but also legal, organisational and infrastructural aspects as well.

The number of treatment technologies (treatment steps) which are available for application in a closed water loop is substantial. The consequence is that for each treatment goal several treatment technologies or several modifications of one type of technology can be considered. At present also the standard wastewater treatment technologies can be applied for closing loop. However it is evident that in future improved and more innovative technologies are required. The main reason is that with increasing attention for sustainable production processes the character of the wastewater and the aim of the treatment process will change. For individual industries the amount of wastewater will become less, coupled at an increase of concentration of pollutants. A typical aspect in closed water loop systems is that, dependent on the type of pollutant, very often low or moderate removal efficiency will be sufficient. The only criteria are that the wastewater is purified to a quality, which is acceptable as process water.

Some type of toxic pollutants which at present may be present in the wastewater will be banned out. This makes the composition of the wastewater less complex so that compounds can more easily be recovered for reuse. Taking these aspects into account, together with other specific aspects related to closed water loop systems, it can be expected that future developments in the treatment technology of wastewater will be focused on

- Anaerobic treatment

- Membrane bioreactors
- Advanced biofilm processes
- Membrane separation processes
- Advanced precipitation processes for recovery of nutrients
- Selective separation processes for recovery of heavy metals
- Advanced oxidation process AOPs
- Physical/chemical process which can be applied at elevated temperature
- Selective adsorption processes
- Advanced processes for demineralisation

Finally it has to be noted that in the development and design of closed loop water systems also attention has to be paid to the implementation route. Successful implementation requires strong involvement of both management and operational staff, also during the design period.

REFERENCES

1. Closing the loop. WQI January/February 1999, p. 38-41.
2. 1st World Water Congress of the International Water Association (IWA), Industrial Wastewater Treatment, Conference Preprint, Book 1. Paris, 3-7 July, 2000.
3. 1st World Water Congress of the International Water Association (IWA), Small Wastewater Treatment Plants, Management of Sludges and Leachates, Conference Preprint, Book 4. Paris, 3-7 July, 2000.
4. 1st World Water Congress of the International Water Association (IWA), Wastewater, Reclamation, Recycling and Reuse, Conference Preprint, Book 8. Paris, 3-7 July, 2000.
5. Asano, T., Planning and implementation of water reuse projects. Wat.Sci.Technol. 24(9): p. 1-10, 1991.
6. Asano, T. and Levine, A.D., Wastewater Reclamation, Recycling and Reuse: An Introduction. Chapter 1 of Wastewater Reclamation and Reuse, ed. Takashi Asano, Vol. 10, Water Quality Management Library, p 1-56, Technomic Publishing, PA., 1998.
7. Baetens, D., Water pinch analysis: minimisation of water and wastewater in the process industry, Chapter 11 in: Water recycling and resource recovery in industry: Analysis, technologies and implementation, Edited by P.Lens et al., IWA publishing, 2002, ISBN: 1 84339 005 1.
8. Ciardelli, G., Capanelli, G. and Bottini, A., Ozone Treatment of Textile Wastewaters for Reuse, in: *Proceedings of the 2nd International Conference on Oxidation Technologies for Water and Wastewater Treatment,* Clausthal-Zellerfeld, Germany, 28-31 May, 2000.
9. Joore, L., Verstraeten, E. and Hooimeijer, A., Competitive Dutch Paper and Board Industry by Closed Water Systems beyond 2000, in: *1999 International Water Conference Towards Closed Water Systems in Papermaking,* Feb 10th, Papendal, Arnhem, TNO, Delft, NL, 1999.
10. Kappen, J. and Wilderer, P.A., Key parameter methodology for increased water recovery in the pulp and paper industry, Chapter 12 in: *Water recycling and resource recovery in*

industry: Analysis, technologies and implementation, Edited by P.Lens et al., IWA publishing, 2002, ISBN: 1 84339 005 1.

11. Kayser, R., Wastewater Treatment by combination of Chemical Oxidation and Biological processes, in: *Proceedings of the International Conference on Oxidation Technologies for Water and Wastewater Treatment,* Clausthal, Germany. Publisher: CUTEC Institut GmbH, 1996.

12. Lazarova, V., Potentials of biotechnolgy in water and resource cycle management, Chapter 18 in: *Water recycling and resource recovery in industry: Analysis, technologies and implementation,* Edited by P.Lens et al., IWA publishing, 2002, ISBN: 1 84339 005 1.

13. Lazarova, V. and Manem, J., Innovative biofilm treatment technologies for water and wastewater treatment, in: *Biofilms II: Process Analysis and Applications,* ed. J.D. Bryers, p. 159-206, Wiley-Liss Inc., 2000.

14. Lens, P.N.L., Vallero, M., Gonzalez-Gil, G., Rebac, S., and Lettinga, G. Environmental protection in industry for sustainable development, Chapter 3 in: *Water recycling and resource recovery in industry: Analysis, technologies and implementation,* Edited by P.Lens et al., IWA publishing, 2002, ISBN: 1 84339 005 1.

15. Levine, A.D., Asano, T., Water reclamation, recycling and reuse in industry, Chapter 2 in: *Water recycling and resource recovery in industry: Analysis, technologies and implementation,* Edited by P.Lens et al., IWA publishing, 2002, ISBN: 1 84339 005 1.

16. Lijmbach, D., J.E. Driver, W. Schipper, Posphorous recycling potentals, Chapter 25 in: *Water recycling and resource recovery in industry: Analysis, technologies and implementation,* Edited by P.Lens et al., IWA publishing, 2002, ISBN: 1 84339 005 1.

17. Maurer, M., Muncke, J. and T.A. Larsen, Technologies for nitrogen recovery and reuse, Chapter 24 in: *Water recycling and resource recovery in industry: Analysis, technologies and implementation,* Edited by P.Lens *et al.* IWA publishing, 2002, ISBN: 1 84339 005 1.

18. Mattioli, D., Malpei, F., Bortone, G. and Rozzi, A., Water minimisation and reuse in the textile industry, Chapter 27 in: *Water recycling and resource recovery in industry: Analysis, technologies and implementation,* Edited by P.Lens et al., IWA publishing, 2002, ISBN: 1 84339 005 1.

19. Mels, A.R. and Teerikangas, E., Physico-chemical wastewater treatment, Chapter 21 in: *Water recycling and resource recovery in industry: Analysis, technologies and implementation,* Edited by P.Lens et al., IWA publishing, 2002, ISBN: 1 84339 005 1.

20. Schuiling, R.D. and Anderade, A. Recovery of struvite from calf manure, *Environmental Technol.,* 20, 1999: p. 756-768.

21. Vogelpohl, A., Advanced oxidation technologies for industrial water reuse, Chapter 22 in: *Water recycling and resource recovery in industry: Analysis, technologies and implementation,* Edited by P.Lens et al., IWA publishing, 2002, ISBN: 1 84339 005 1.

22. Worp, J.J.M. van der, Sustainable water management in industry, Chapter 1 in: *Water recycling and resource recovery in industry: Analysis, technologies and implementation,* Edited by P.Lens et al., IWA publishing, 2002, ISBN: 1 84339 005 1.

OPTIMISATION OF REACTOR TECHNOLOGY FOR SELECTIVE OXIDATION OF TOXIC ORGANIC POLLUTANTS IN WASTEWATER BY OZONE

Wim H. Rulkens, Harry Bruning, Marc A. Boncz
Wageningen University and Research Centre, Sub-department of Environmental Technology, PO. Box 8129, 6700 EV Wageningen, The Netherlands. Tel. +31 317 483339, Fax +31 317 482108, E-mail Wim.rulkens@algemeen.mt.wag-ur.nl

Abstract: Oxidation of toxic pollutants by ozone is often required as a process step preceding an anaerobic or aerobic treatment that is focused on the removal of the non-toxic organic pollutants from the wastewater. The costs of such a pretreatment step strongly depends on the amount of ozone required for the oxidation process and therefore on the selectivity of the ozone oxidation process. Factors governing this selectivity are reaction kinetics of the oxidation process of toxic and non-toxic pollutants with ozone, reactor type, ozone supply to the reactor, temperature, and mass transfer limitations. By means of a mathematical model the influence of these process parameters on the selectivity is calculated. The results can be used for an optimal design of the treatment process.

Key words: Ozone, selective oxidation, toxic pollutants, wastewater, mass transfer, limitations, reactor design

1.   INTRODUCTION

The past two decades have shown an increasing interest in advanced physical-chemical processes for wastewater treatment[12]. There are several reasons for this increasing interest. First, with conventional aerobic and anaerobic biological wastewater treatment technologies many industrial wastewater streams can not be treated to a high effluent quality[4]. Physico-chemical wastewater treatment systems are often considered as an appropriate alternative or can be applied as an additional treatment system.[6, 9, 10.] When the presence of strongly toxic organic compounds in the wastewater will disturb the biological treatment process completely, this physico-chemical treatment step has to be applied as a pretreatment process[10]. Another reason for the increased interest in the physico-chemical processes is that the water management approach within the industry focuses more and more on closed loop systems, including the reuse of treated wastewater streams[14]. However, very often a biological process or a combination of biological processes can not achieve the required effluent quality.

To obtain a water quality that is suitable for reuse it will often be necessary to include a physico-chemical treatment step. A large variety of physico-chemical processes or process steps exist which can be applied in practice in combination with biological processes. Well known processes are: membrane filtration, precipitation, flotation, adsorption, chemical oxidation, electrodialysis and ion exchange. Each process has its own specific field of application. In the case of strongly toxic soluble pollutants, like chlorinated aromatics, phenols, dyes, pesticides, etc., Advanced Oxidation Processes (AOPs) may be suitable destruction techniques. In general these AOPs are rather expensive due to the amounts of chemicals that are required and/or due to the costs of energy needed for these processes. Especially in the case of waste waters containing small amounts of such pollutants in addition to large quantities of other easily oxidizable and biodegradable compounds, the application of only a physico-chemical treatment step to completely remove all pollutants will in general not be economically feasible. In that respect there is a strong need for oxidation processes which can selectively oxidize toxic pollutants, prior to a biological treatment step. A possible pretreatment technique is the application of ozone.

This paper deals with the selective (partial) oxidation of toxic compounds by ozone in wastewater that also contains large amounts of non-toxic easily biodegradable pollutants. For a technically and economically optimal reactor design three aspects are relevant: reaction kinetics, reactor type and process conditions. The reactor design for the oxidation process can be optimized regarding both the selectivity of the reaction and the use of ozone when the reaction kinetics of the oxidation of the waste water constituents are known. Data regarding these reaction kinetics can be taken from literature[1]. Reactor type and process conditions applied may strongly influence the amounts of ozone necessary for the selective oxidation of toxic pollutants. This is a factor, which is rarely discussed in literature dealing with ozone oxidation. This paper will discuss the effect of two different reactor types on the amount of ozone necessary for selective oxidation. These reactors are the continuous flow stirred tank reactor (CFSTR) and the plug flow reactor (PFR). The oxidation processes in these reactors are described by means of simplified mechanistic models. Calculations are made on the effect of process conditions on the effectiveness and efficiency. Another aspect which will be discussed is the possible effect of mass transfer limitations on the selectivity of the oxidation process. In literature very little attention is paid to this process parameter. The final aim of this paper is to present a model in which the effect of the various relevant process parameters which may influence the selectivity and the consumption of ozone is illustrated. The model will allow for optimising the process design in such a way that an organic pollutant will be oxidised by ozone selectively utilising a minimum amount of ozone.

2.   REACTION RATE CONSTANTS

In an ozonisation reactor three reaction types can occur: 1. reactions in which organic compounds are directly oxidized by ozone, 2. reactions in which organic compounds are oxidized by radicals, mainly hydroxyl radicals, and 3. reactions leading to the decomposition of ozone into oxygen, following a pathway in which inorganic radicals are formed as intermediates. The process design can be optimized when data concerning the relevant reaction kinetics are available. These include both the reactivity of the organic compounds with ozone and the kinetics of the decay of ozone in water. Numerous reaction rate constants can be found in literature for the first two categories of reactions[2, 3, 5, 7, 8, 11]. From an overview of a large number of reaction rates constants, both measured and obtained from literature, it is clear that the oxidation with ozone will be far more selective than the oxidation with hydroxyl radicals derived from this oxidant. Where the reaction rate constants of the reaction with ozone differ by 10 orders of magnitude, the reaction rate constants for the radical reaction only differ by 3 orders of magnitude. When studying these data in more detail a few trends can be observed: aliphatic alcohols and acids hardly show any direct reactivity with ozone, aromatic acids aren't very reactive as well, but phenols and amines (both aromatic and non-aromatic) react with ozone at a very high rate, and will most likely be oxidized by a direct reaction in all situations, even when the reaction conditions are such that a relatively high concentration of radicals should be produced. When these fast reacting compounds are present, the process will most likely be mass-transfer controlled, although this will not always significantly affect the selectivity of the process, unless steric factors play a role in the system. Here we will only consider the situations where the direct reaction mechanism (oxidation by the ozone molecule) is the predominant degradation route and degradation by the hydroxyl radical can be neglected, a realistic approximation when considering highly concentrated wastewaters.

3.   MODEL REACTOR SYSTEMS: STARTING POINTS AND BOUNDARY CONDITIONS

The oxidation process in a wastewater containing small amounts of a toxic component A and a high concentration of a non-toxic easily biodegradable pollutant B involves three types of reactions:

$$A + z_A O_3 \xrightarrow{k_A} P_A \qquad\qquad r_A = -k_A C_{O_3} C_A \quad (1)$$

$$B + z_B O_3 \xrightarrow{k_B} P_B \qquad\qquad r_B = -k_B C_{O_3} C_B \quad (2)$$

$$2 O_3 \xrightarrow{k_O} 3 O_2 \qquad\qquad r_O = -k_O C_{O_3} \quad\;\; (3)$$

in which z_A and z_B are stoichiometric coefficients, C_A, C_B and C_{O_3} are the concentrations of compound A, B and ozone respectively and P_A and P_B are the oxidation products. r_A, r_B and r_O are the reaction rates of the three reactions, in which k_A, k_B and k_O are the reaction rate constants. The reaction is selective with $k_A > k_B$. The third reaction, the spontaneous decay of ozone, is assumed to be first order, according to literature. For both reactor systems, it is assumed that the ozone is fed as a gas. It is further assumed that component A has to be converted to reach a certain effluent concentration C_{Ae}, defined by the maximum allowed concentration of this compound in the effluent that is fed to a bioreactor system.

4.   REACTION IN CONTINUOUS FLOW STIRRED TANK REACTOR AND IN A PLUG FLOW REACTOR

4.1   4.1 Continuous Flow Stirred Tank Reactor (CFSTR)

We first consider the continuous flow stirred tank reactor (CFSTR). A schematic presentation of the continuous flow stirred tank reactor is given in Figure 1. It is assumed that no mass transfer limitations exist regarding the supply of ozone. The accuracy of this assumption depends on the way ozone is supplied to the system and the reaction rate constants of the components involved.

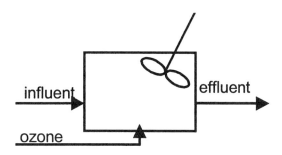

Figure 1. Schematic presentation of a CFSTR

The total ozone consumption per unit time, Φ_{ozone} is given by:

$$\Phi_{ozone} = z_A Q (C_{Ai} - C_{Ae}) + z_B Q (C_{Bi} - C_{Be}) + (Q + k_O V) C_{O_3e} \qquad (4)$$

in which Q is the volume flow through the reactor, C_{Ai}, C_{Ae}, C_{Bi}, C_{Be}, are the influent and effluent concentrations of component A and component B respectively. C_{O_3e} is the effluent concentration of ozone and V is the volume of the reactor. Φ_{ozone} can be expressed in kinetic constants and influent composition. From mass balances over the reactor and the conversion rate equations (1, 2, 3) it can be calculated that for a given residence time $\theta = V/Q$, the conversion of component A and component B are given by:

$$\frac{C_{Ae}}{C_{Ai}} = \varepsilon = \frac{1}{1 + k_A C_{O_3e} \theta} \qquad (5)$$

$$\frac{C_{Be}}{C_{Bi}} = \frac{1}{1 + k_B C_{O_3 e} \theta} \qquad (6)$$

The value of the conversion factor ε is determined by the required efficiency of the conversion of component A. By substituting equations (5,6) into equation (4) we find for the ratio of the total ozone consumption to the amount of ozone needed for the oxidation of component A, defined as the Ozone consumption Factor (OF):

$$OF = \frac{\Phi_{ozone}}{Q z_A (C_{Ai} - C_{Ae})} = 1 + \frac{\frac{z_B C_{Bi}}{z_A C_{Ai}}}{1 + \varepsilon \left(\frac{k_A}{k_B} - 1 \right)} + \frac{k_O}{z_A k_A C_{Ai} \varepsilon} + \frac{1}{z_A k_A C_{Ai} \varepsilon \theta} \qquad (7)$$

With given flow Q and influent concentrations C_{Ai} and C_{Bi}, and the required efficiency ε the reactor volume and ozone consumption can be calculated from this equation.

At the right hand side of this equation we have four terms dealing with four different way of ozone consumption. The first term represents the relative amount of ozone dealing with the ozone consumption by component A. The second term gives the relative ozone consumption due to the reaction of component B with ozone. It is clear from equation (7) that especially the ratio of the inlet concentrations, C_{Bi} and C_{Ai}, are an important factor. Even at high values of k_A and k_B this ratio is crucial because the value of the conversion factor ε is low. The third term gives the relative ozone consumption due to the decay of ozone. High losses can be expected if we have to deal with a high conversion factor ε and if the values of k_0 and k_a are of the same order of magnitude. The fourth term finally shows the consumption of ozone due to ozone loss with the effluent.

From equation (7) is also clear that the volume of the reactor V has only an effect on the ozone loss with the effluent. An increase in reactor volume results in an increase in residence time θ and therefore in a decrease of the ozone loss. This means that large reactors are favourable. Also with respect to ozone supply a large reactor has advantages. However it has to be noted that large reactors require more mixing energy and are also more expensive.

4.2   Plug Flow Reactor (PFR).

A schematic presentation of a plug flow reactor is given in Figure 3. The length of the reactor is L. Ozone is supplied to the reactor over the entire reactor length, with exception of the last part of the reactor. In a plug flow reactor the process can be further optimised by taking into account the spatial distribution of ozone.

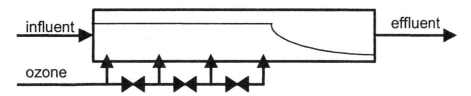

Figure 2. Schematic presentation of a PFR. The line in the reactor indicates the ozone concentration as a function of the reactor length.

In the case of a plug flow reactor (PFR) the ozone consumption rate Φ_{ozone} can be given by an equation comparable to that for the CFSTR:

$$\Phi_{ozone} = Qz_A\left(C_{Ai} - C_{Ae}\right) + Qz_B\left(C_{Bi} - C_{Be}\right) + Vk_O\left\langle C_{O_3}\right\rangle + QC_{O_3e} \quad (8)$$

Here $\left\langle C_{O_3}\right\rangle$ is the average ozone concentration in the PFR. From the

conversion rate equations and mass balances for component A and B, it can be derived that:

$$u\frac{dC_A}{dx} + k_A C_{O_3} C_A = 0 \quad\quad\quad (9)$$

$$u\frac{dC_B}{dx} + k_B C_{O_3} C_B = 0 \quad\quad\quad (10)$$

in which u is the linear flow velocity. Equation (9) can be rewritten as:

$$u\frac{dC_A}{C_A} = -k_A C_{O_3} dx \quad\quad\quad (11)$$

Integration of this equation between x = 0 (C_A = C_{Ai}) and x = L (C_A = C_{Ae}) results in:

$$\langle C_{O_3} \rangle = \frac{1}{k_A \theta} \ln\left(\frac{1}{\varepsilon}\right)$$

(12)

Combining of equation (9) and (10) results in:

$$\frac{dC_B}{dC_A} = \frac{k_A}{k_B} \frac{C_B}{C_A}$$

(13)

or

$$\frac{dC_B}{C_B} = \frac{k_A}{k_B} \frac{dC_A}{C_A}$$

(14)

Integration of equation (14) results in:

$$\frac{C_{Be}}{C_{Bi}} = \varepsilon^{k_B/k_A}$$

(15)

From equation (12) it can be concluded that the required residence time θ to achieve a certain value of ε is inversely proportional to the average concentration of ozone present in the reactor. The value of the required residence time and therefore the required volume V of the reactor is completely independent of the distribution of ozone over the reactor. This also holds for the conversion of component A and component B.

Substitution of equation (12) and (15) into equation (8) yields the ozone consumption factor (OF) for the PFR:

$$OF = \frac{\Phi_{ozone}}{Qz_A(C_{Ai} - C_{Ae})} = 1 + \frac{z_B C_{Bi}}{z_A C_{Ai}}\left(\frac{1 - \varepsilon^{k_B/k_A}}{1 - \varepsilon}\right) + \frac{k_O \ln\left(\frac{1}{\varepsilon}\right)}{z_A k_A C_{Ai}(1 - \varepsilon)} + \frac{C_{O_3 e}}{z_A C_{Ai}(1 - \varepsilon)}$$

(16)

From this analysis we see that the selectivity is independent of the spatial distribution of the ozone concentration. Minimal ozone dosage can be obtained when $C_{O_{3e}} = 0$. This can be achieved by dividing the PFR into two zones (see Figure 2): a first zone, to which the influent is fed and in which ozone is introduced, and a second zone without introduction of ozone in which the ozone concentration will decrease to $C_{O_{3e}} = 0$.

For a given conversion factor ε and a given average ozone concentration, $\langle C_{O_3} \rangle$, required for the conversion process, the value of $C_{O_{3e}}$ can be calculated from a mass balance. For the second zone, starting at x = L_1, this is the zone in which no supply of ozone occurs, the following mass balance for ozone can be derived:

$$u\frac{dC_{O_3}}{dx} + k_0 C_{O_3} + k_A C_{O_3} + k_A C_{O_3} C_A + k_B C_{O_3} C_B = 0 \qquad (17)$$

The boundary conditions which have to be used are:

$$x=L_1 \qquad C_{O_3} = C_{O_3 L_1}, \quad C_A = C_{AL_1}, \quad C_B = C_{BL_1}$$

$$x=L \qquad C_{O_3} = C_{O_{3e}}, \quad C_A = C_{Ae}, \quad C_B = C_{Be}$$

Together with equation (9) and (10) equation (17) can be solved numerically. The numerical solution will not be discussed further.

4.3   Comparison of CSTFR and PFR

The consumption of ozone by component B is lower when using a PFR instead of a CFSTR, as can be calculated from the second terms in the right hand sides of equations (4) and (8). In Figure 3 the ratio of the ozone consumption by the conversion of component B in a PFR and a CFSTR is given in dependence on the ratio k_A/k_B and ε. When a high conversion of component A is desired ($\varepsilon \ll 1$) and the k_A and k_B differ by more than one order of magnitude, which is mostly the case when only oxidation by molecular ozone is considered the ozone losses due to the conversion of component B can be reduced by more than 90%.

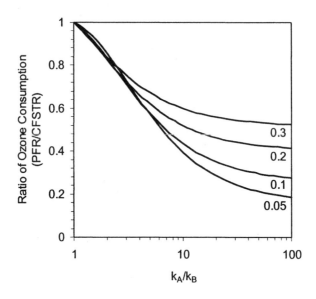

Figure 3. Ratio of the ozone consumption by the conversion of component B (second term in right hand side of equations (7) and (16)) in a PFR and a CFSTR as a function of kA/kB. Parameter is □ = CAe/CAi.

The absence of mixing (PFR) leads to an optimal ozone usage. If it is assumed that in the PFR the concentration at the exit approaches 0 (optimal conditions) then the amount of ozone self-decomposition in the PFR can be compared to that in the CFSTR using the third term in the right-hand side of the equations (4) and (8). From this comparison it is clear that in the CFSTR more ozone will be lost due to self-decomposition than in the PFR.

The loss of ozone due to the decay of ozone is always lower in a PFR than in a CSTR. This is clearly shown if we look to the ratio of the relevant terms in equation (7) and equation (16). This ratio, defined as, R_D,

$$R_D = \frac{\varepsilon \ln\left(\frac{1}{\varepsilon}\right)}{(1-\varepsilon)} \tag{18}$$

is graphically shown in Figure 4. From the graph it is clear that especially at low values of ε, order of magnitude 0.1 to 0.01, R_D strongly increases with decreasing ε.

Minimal losses are obtained in a PFR, while the losses are maximal for a CFSTR. From equation (4), it can be derived that, in the case of a CFSTR, the ratio of the ozone losses by decomposition and losses in the effluent is represented by $k_O\varepsilon$. Because in most cases $k_O \ll 1$, ozone losses in the effluent of a CFSTR are inevitably high.

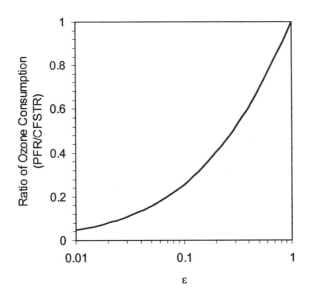

Figure 4. The ozone consumption ratio for the spontaneous decay of ozone (third term in right hand side of equations (7) and (16)) as a function of $\square$ = CAe/CAi

All three terms which contribute to ozone losses are minimal in a PFR. The relative importance and the absolute values of these terms and the ozone consumption factor can be calculated from equations (7) and (16), provided the reaction rate constants and the reaction conditions are known.

It is also interesting to compare the ozone losses due to the presence of ozone in the effluent. To that aim we consider the ideal situation that in both reactors the ozone concentration is distributed equally over the reactor volume. Further we assume that both reactor systems have the same average residence time, θ. The average concentration of ozone in the CFSTR and PFR is given by equation (5) and (12) respectively. From these two equations it can be derived that the ratio of the average ozone concentration in the PFR and the CFSTR, R_e is given by:

$$R_e = \frac{\varepsilon \ln\left(\frac{1}{\varepsilon}\right)}{(1-\varepsilon)} \tag{19}$$

This is the same relationship as already derived for the decay of ozone, equation (18). From this it is clear that in case of a PFR generally the loss of ozone with the effluent is lower than in case of a CFSTR.

5.   MASS TRANSFER LIMITATIONS

5.1   Reactor systems for supply of ozone

Several reactor systems are available to transfer ozone from a gas phase to a water phase. The most important systems are:
- The bubble column
- The stirred tank reactor
- The air lift reactor
- The packed column
- The tray column
- The spray column
- The venturi reactor

The system mostly applied in practice for supply of ozone is the bubble column and the stirred tank reactor. With these reactor systems it is always possible to set up the complete reactor modification as a plug flow reactor, a continuous flow single stirred tank reactor or a cascade of stirred tank reactors.

5.2   Mass transfer without chemical reaction

Mass transfer of ozone from the gas phase to the liquid phase strongly depends on the reactor system and the process conditions. Two characteristic factors are in general important: the overall mass transfer coefficient between gas phase and water phase, k_{tot}, and the specific surface area available for mass transfer, A_m.

The mass transfer rate of ozone, N_0, is given by the equation

$$N_0 = k_{tot} \, A_m \left(m C_{gO_3} - C_{0_3} \right) \tag{20}$$

where

k_{tot} = overall mass transfer coefficient for ozone

C_{O_3} = ozone concentration in the water phase

C_{gO_3} =   ozone concentration in the gas phase

m   = the ratio between the ozone concentration in the
water phase and the ozone concentration in the gas
phase which is in equilibrium with the
concentration of ozone in the water phase

To transfer ozone to a water phase very often fine bubble columns or
stirred or agitated tank reactors are applied. The bubbles in these reactors are
small. The value of m for ozone is relatively high. Furthermore the diffusion
coefficient of ozone in the gas phase is very high compared with the
diffusion coefficient of ozone in the water phase. These three reasons lead to
the conclusion that in bubble columns and stirred tank reactors the resistance
to mass transfer in the gas phase can be neglected. Equation (20) can than be
simplified to:

$$N_0 = k_{liq} \left(mC_{gO_3} - C_{O_3} \right)$$ (21)

where

k_{liq} = the liquid mass transfer coefficient for ozone

Similar to k_{tot}, k_{liq} strongly depends on process conditions such as
- Size of the gas bubbles d_b
- Diffusivity of ozone in the liquid phase D_0
- Relative velocity of the gas bubbles v_r
- Viscosity of the liquid phase η
- Density of the liquid phase ρ

The latter three factors are only relevant for the mass transfer if the
Reynolds number ($Re = \rho \, v_r \, d_b / \eta$) of the liquid flow around the particle is
larger than 1. The size of the gas bubbles depends on liquid properties such
as temperature, surface tension and viscosity but also on the dissipated
power. If we have to deal with small gas bubbles in a bubble column than we
can consider the gas bubbles as rigid. The mass transfer coefficient k_{liq} is
then given by the equation:

$$Sh = \frac{k_{liq}d_b}{D_o} = 2 \tag{22}$$

where *Sh* is the Sherwood number and D_0 the diffusivity of ozone in the liquid.

The value of A_m strongly depends on the size distribution of the bubble diameter in the system and the volume fraction of the bubbles. These factors are strongly dependent not only on the liquid properties, but also on the gas/liquid flow ration and the energy dissipation. Opposite to the situation with respect to k_{liq} it is not possible to calculate A_m with theoretical equations. This is the reason that in practise almost always the product of k_{liq} and A_m is considered. This product can easily be determined experimentally.

For bubble columns the value of $k_{liq} A_m$ strongly depends on the superficial gas velocity. For the stirred tank reactor gas sparged is immediately sucked into the gas cavities behind the stirred blades. The value of k_{liq} depends strongly on both the superficial gas velocity, the pressure at the stirrer level and the liquid volume.

5.3   Mass transfer with chemical reaction

As already has been mentioned mass transfer of ozone from the gas phase to the liquid phase may be enhanced by the chemical reactions of ozone with components A and B and by the decay of ozone. The effect of this enhancement in mass transfer on the selectivity will be discussed now semi-quantitatively[13]. To that aim we consider a gas phase in contact with a liquid phase. The liquid phase consists of a thin stagnant film at the interface with the gas phase, and a liquid bulk phase. We assume that the ozone is completely converted in the stagnant liquid film. This is for example the case if we have to deal with a high reaction rate constant and a relatively high concentration of one of the pollutants in the liquid film. Figure 5 gives a schematically presentation of this situation.

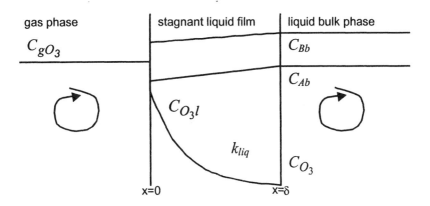

Figure 5. Transport through liquid film with chemical reaction.

Now we consider the situation that in the liquid film the ozone is completely converted by reaction with component A and component B. The decay of ozone is neglected. In the liquid film transport of ozone and component A and B takes place only by diffusion. From a mass balance the following equation for ozone can be derived.

$$D_0 \frac{d^2 C_{O_3}}{d x^2} - k_A C_{O_3} C_A - k_B C_{O_3} C_B = 0 \qquad (23)$$

where x is the co-ordinate in the direction of the transport. For the components A and B similar equations can be derived.

$$D_A \frac{d C_A^2}{d x^2} - k_A C_{O_3} C_A = 0 \qquad (24)$$

where D_A is the diffusivity of component A.

$$D_B \frac{d^2 C_B}{d x^2} - k_B C_{O_3} C_B = 0 \qquad (25)$$

where D_B is the diffusivity of component B

We consider now the situation that component A and B are not volatile and that the volume of the stagnant film is small compared with the bulk of the liquid. This means that the bulk concentration of component A and B can be assumed to be constant. In fact we consider a quasi stationary process. Then the boundary conditions of equations (23), (24) and (25) are:

$$x = 0: \qquad C_{0_3} = m C_{g0_3} = C_{0_3 l}, \quad D_A \frac{d C_A}{d x} = 0, \quad D_B \frac{d C_B}{dx} = 0 \qquad (26)$$

$$x = \delta: \qquad C_{0_3} = 0, \quad C_A = C_{Ab}, \quad C_B = C_{Bb} \qquad (27)$$

where C_{Ab} and C_{Bb} are the bulk concentrations of component A and B respectively and δ is the thickness of the stagnant liquid film.

Now we define three dimensionless factors: The (dimensionless) Hatta number Ha_A related to ozone and component A:

$$Ha_A = \frac{\left(D_0 k_A C_{Ab}\right)^{1/2}}{k_{liq}} \qquad (28)$$

The dimensionless Hatta number Ha_B related to ozone and component B:

$$Ha_B = \frac{\left(D_0 k_B C_{Bb}\right)^{1/2}}{k_{liq}} \qquad (29)$$

The dimensionless factor ϕ_A:

$$\phi_A = \frac{D_A C_{Ab}}{D_{0_3} C_{0_3 l}} \qquad (30)$$

where $C_{0_3 l}$ is the concentration of ozone in the liquid at the gas liquid interface.

The dimensionless factor ϕ_B:

$$\phi_B = \frac{D_B \, C_{Bb}}{D_{O_3} \, C_{O_3 l}} \tag{31}$$

The equations (23) to (27) can only be solved numerically. However such a numerical solution gives less insight in the factors governing the transport and conversion processes. Therefore we consider another approach. In this approach, the transport and conversion of component A are calculated under the assumption that no reaction of ozone with component B takes place. The enhancement factor for mass transfer of ozone, E_A, can now be given by the equation:

$$E_A = \frac{- D_O \left. \dfrac{d \, C_{O_3}}{d \, x} \right|_{x=0}}{k_{liq} \, C_{O_3 l}} \tag{32}$$

The relationship between E_A, Ha_A and ϕ_A can be calculated numerically and is given in Figure 6.

In general the concentration of component A (the toxic pollutant) is low compared to the concentration of ozone. That means that the value of ϕ_A is smaller than or in the order of 1. The enhancement factor for mass transfer can then be neglected. There is also no influence of the Hatta number Ha_A. In a similar way we can calculate the transport and diffusion of component B under the assumption that no reaction of ozone with component A takes place. The enhancement factor for mass transfer of ozone, E_B, can then be given by the equation

$$E_B = \frac{- D_O \left. \dfrac{d \, C_{O_3}}{d \, x} \right|_{x=0}}{k_{liq} \, C_{O_3 l}} \tag{33}$$

The relationship between E_B, Ha_B and ϕ_B is also given by Figure 6.

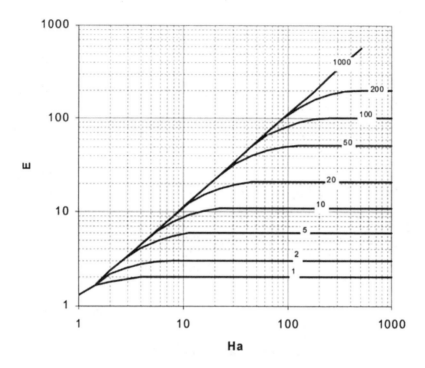

Figure 6. Enhancement factor E as a function of Ha. Parameter is ϕ.

In general the concentration of component B (the non-toxic pollutant) is higher than the concentration of ozone. That means that ϕ_B is much higher than 1. There is also a strong influence of the Hatta number, in this case, Ha_B. It can therefore be concluded that at high values of Ha_B, the enhancement of ozone mass transfer due to the reaction of ozone with component B may be substantial.

From the above it can be concluded that only the reaction with component B may enhance mass transfer of ozone substantially. And only if the Hatta number Ha_B is much higher than 1. Therefore it can be expected that whenever we have to deal with an enhancement of mass transfer due to chemical reactions, this influences the selectivity of the oxidation process in a negative way. The factor which has to be considered in this respect is the Hatta number for the reaction of ozone with component B (equation 29). Ha_B increases with increasing value of k_B and C_{Bb} and with decreasing value of the mass transfer coefficient for ozone, k_{liq}.

6.   CONCLUSIONS

Many wastewater flows in industry can not be treated by standard aerobic or anaerobic treatment methods due to the presence of relatively low concentration of toxic pollutants. Ozone can be used as a pretreatment step for the selective oxidation of these toxic pollutants. Due to the high costs of ozone it is important to minimise the loss of ozone due to reaction of ozone with non-toxic easily biodegradable compounds, ozone decay and discharge of ozone with the effluent from the ozone reactor. By means of a mathematical model, set up for a plug flow reactor and a continuos flow stirred tank reactor, it is possible to calculate more quantitatively the efficiency of the ozone use, independent of reaction kinetics, mass transfer rates of ozone and reactor type. The model predicts that the oxidation process is most efficiently realised by application of a plug flow reactor instead of a continuous flow stirred tank reactor.

REFERENCES

1. Boncz, M.A., et al., Substituent effects in the Advanced Oxidation of Aromatic Compounds, in: *Proceedings of the International Regional Conference of the IOA,* Poitiers, 1998, France.
2. De Laat, J., P. Maouala Makata, and M. Doré, Rate constants for reactions of ozone and hydroxyl radicals with several phenyl-ureas and acetamides, *Environmental Technology,* 17(7), 1996, p. 707-16.

COMPARISON BETWEEN SEQUENCING BATCH AND CONTINUOUS FLOW ACTIVATED SLUDGE SYSTEMS

Janusz A. Tomaszek
Dept. of Environmental & Chemistry Engineering, Rzeszów University of Technology, 2 W Pola Street, 35-959 Rzeszów, Poland, E-mail: tomaszek@prz.rzeszow.pl

Abstract: After introducing fundamentals of periodic processes occurring in sequencing batch reactor (SBR) a comparison of SBR to plug flow reactor (PFR) and completely mixed flow reactor (CMFR) was made.

Key words: Activated sludge systems, SBR, continuous flows, CNP removal

1. INTRODUCTION

During a few last years some papers have reported that SBR-s outperform continuous flow activated sludge systems. Superiority of performance of SBR relates to simplicity of their operation, production of non-bulking sludge, simultaneous nitrification and denitrification, production of less sludge using less energy, the absence of scum problems, higher rank of phosphorus removal (Goronszy, Rigel, 1991). While there are many application for which the SBR system is highly suited, for others it is rather groundless.

The aim of this paper was to compare periodic and continuous processes showing their advantages and disadvantages.

2. DESCRIPTION OF SBR PROCESS

Sequencing batch reactor technology has been developed on a scientific assumption that periodic exposure of microorganisms to defined process conditions is effectively achieved in a fed-batch system in which exposure time, frequency of exposure and amplitude of the respective concentrations can be set independently of any inflow conditions.

The SBR process is characterized by a series of process phases, each lasting for a defined period during each cycle. The total cycle time, t_C is the duration corresponding to the sum of five phases: fill, t_F, react, t_R, settle, t_S, decant, t_D, idle, t_I.

$$t_C = t_F + t_R + t_S + t_D + t_I \qquad (1)$$

The cycle begins with fill and terminates at the end of the idle phase. In the fill phase the wastewater is fed into the reactor on the settled activated sludge remained from the previous cycle. After fill, additional time is given in the react phase for biological conversion. The microorganisms are allowed to settle to the reactor bottom in the settle phase. The clear supernatant is discharged in the draw phase. The reactor is left idle during the idle phase. In addition to the descriptions above, fill and react can have several sub-phases based on the energy input to the system, which results in various aeration and mixing operation strategies. The volume of wastewater introduced into the reactor is ΔV_F. It is added to the volume of water and sludge that remains in reactor at the end of the past cycle (V_O). At the end of the fill phase the reactor contains $V_{max} = V_O + \Delta V_F$. After removing of excess sludge, ΔV_w and discharge of the treated supernatant, ΔV_D , the reactor is available to receive a new supply of wastewater.

SBR process is basically characterized by the following sets of parameters: t_i – time for the ith phase, t_C – total time of one cycle ($t_C = \Sigma t_i$), FTR – fill time ratio, t_F/t_C, VER – volumetric exchange ratio, $\Delta V_F/V_{max}$, HRT (or τ) – hydraulic residence time, $n V_{max} Q^{-1}$, where n – the number of tanks, V_{max}, - total liquid volume of the reactor, Q – volumetric flow rate of the influent to the treatment plant for each of the tanks, $HRT_i = t_C VER^{-1}$. In addition, process parameters apply that are typical for activated sludge or biofilm systems i.e. sludge age and sludge loading. Because the fill phase is usually only a fraction of the cycle, it is necessary to provide more than one reactor to handle a continuous inflow of wastewater if some temporary influent storage volume is not available.

Integrate nutrient removal in SBR is successfully achieved similarly as in continuous flow systems. It requires multiple fill and react phases during one cycle, with successive periods of aeration and anaerobic or anoxic mixing. It is claimed (Artan et al. 2001), that no biological conversion is assumed to occur during settle, draw and idle phases. The processes take place during the process time (t_P) which corresponds to the sum of fill and react phases unless a static fill is used.

$$t_P = t_F + t_R \tag{2}$$

In the static fill phase, the wastewater is fed into the reactor in a very short time, and it is not mixed with the reactor contents. Therefore, for the static fill the process time (t_P) corresponds to the react phase:

$$t_P = t_R \tag{3}$$

In nutrient removal SBR systems the process phase (t_P) consists of aerated period (t_A) and mixed period (t_M) which can be anoxic (t_{DN}) and anaerobic (t_{AN}) depending on presence of nitrate. Hence:

$$t_P = t_C - (t_S + t_D + t_O) = t_M + t_A = t_{AN} + t_{DN} + t_A \tag{4}$$

According to the German Association for the Water Environment, ATV (Teichgräber B., 1998 and Teichgräber et al. 2001) to calculate t_P from the whole cycle t_C biological phosphorus removal phase t_{Bio-P} should be taking away.

$$t_P = t_C - t_{Bio-P} \tag{5}$$

Because:

$$t_{Bio-P} = t_{AN} \tag{6}$$

The biomass is not effective throughout the whole cycle therefore, the process time (t_P) is therefore the effective duration (t_E) with respect to heterotrophic growth and endogenous respiration (Artan, 2001).

$$t_E = t_C - t_{AN} = t_{DN} + t_A \tag{7}$$

The simplified account of SBR process sequence which is given by equation (1) can differentiate sub-phases depending on static fill, mixed fill without forced aeration allows either anoxic or aerobic reactions, aerated fill allows simultaneous anoxic and aerobic reactions, mixed react without forced aeration, aerated react.

3.   COMPARISON OF THE SYSTEMS

Depending on t_F/t_C ratio, SBR operation can be compared with plug flow reactor (PFR) and completely mixed flow reactor (CMFR), (Weber & DiGiano, 1995). In table 1 mass balance equations for SBR and continuous flow system are compared in which:

C_S – concentration of soluble substrate in reactor S,

$C_{S\,in}$ – concentration of soluble substrate in the effluent,

$r_{s,v}$ – net volumetric rate of formation of substrate, g/m^3h,

A – cross-sectional area of PFR tank,

x – longitudinal distance from the inlet of the PRF.

Dump fill, instantaneous addition of waster water into the reactor is rarely used in the field but is implemented practically by including a static fill in the operating strategy. The mathematical representation for the SBR with dump fill is the same as that for the plug reactor at steady state, where the hydraulic residence time in the PFR compares to the time for react in the SBR.

The primary difference between the PRF and the SBR during react is that the equivalent to true plug flow conditions can be established in an SBA but cannot be achieved in a single activated sludge tank because of the dispersion resulting from the aeration system. However, a cascade of CMFRs consisting three or four tanks in series can be considered as a suitable approximation (Wilderer et al., 2001).

The mass balance equation for the SBR with slow fill resembles that of unsteady-state CMFR with variable volume. As originally conceived, SBR operation includes a react period after fill. Thus, a slow fill system s represented by a CMFR followed by a PFR, the minimum volume configuration for an activated sludge system capable of achieving the desired overall treatment performance (Irvine and Ketchum, 1989).

Table 1. Selected mass balance equations (after Wilderer P.A., et al., (2001).

(a) SBR with dump fill (for react only):	(b) PRF ,
Initial condition:	which at steady state is:
$Cs(0) = Cs_{in}$ VER + $Cs\,(t_C)\,(1 - VER)$	Initial condition:
where t_C is the time for one cycle	$Cs(0) = [Cs_{,in} + Cs\,(\tau_c)\alpha]\,(1 + \alpha)^{-1}$,
Assumes no substrate conversion after react	where t_C is the flow-through time given by:
	HRT/$(1-\alpha)$; assumes constant flow
c) SBR with slow fill (during fill only)	(d) CMFR.
Initial condition: $C_S(0) = C_S(t_C)$	Initial condition: known
Assumes constant flow and either mixed or aerated fill	Assumes constant volume

In Figure 1, (after Wilderer et al., 2001) two SBR cycles are shown, each with its corresponding continuous flow system. The recirculation of sludge and treated water in the PRF system is comparable to the water remaining in the SBR after draw (V_0).
1. By comparing the SBR with dump fill with that for the PFR in Table 1 it can be seen that: VER = $1/(1+\alpha)$, if the two initial conditions are identical, where ; α – is a recycle ratio.

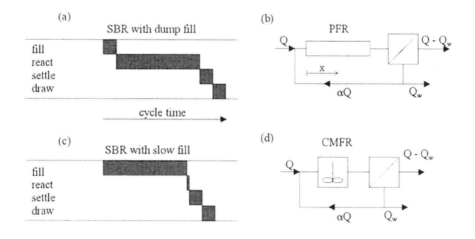

Figure 1. Comparison of SBR reactors to continuous flow systems.

2. In Figure 2, (after Wilderer et al., 2001) relation between volumetric exchange ratio in the activated sludge SBR and the recycle ratio in a continuous flow system is shown. Microorganisms in a controlled unsteady-state activated sludge system are periodically subjected to a series of different environmental conditions. Because the continuous flow tank–in–series cascade and the SBR system are controlled unsteady- state systems, one might expect that the microbial community developed in either would have the same or similar physical and biological properties. This is in fact true. Both can be designed and operated to remove nutrients. However, in practice both systems have developed specific niches. SBR has the advantage of more flexible operation and ideal plug flow characteristics. Because treatment proceeds with time in one reactor, process control is conveniently executed by using input from on-line analysers (for dissolved oxygen, pH, and ammonia). In contrast, the continuous flow system can operate simply, without the aid of computers or timers. The mainstream of wastewater passes through the continuous

flow plant by gravity with the bulk of the pumping being required for the recycling of sludge. SBR systems do not require recycle because sedimentation takes place under quiescent conditions within the biological reactor itself. Both the continuous flow and SBR systems are designed to handle easily a wide range of highly variable influent conditions such as the effluent of combined sewers. When the hydraulic loading increases, the hydraulic retention time in the various tanks of the continuous flow plant (reactors and clarifiers) automatically decreases. The relationship between hydraulic loading and retention time is also applied to the SBR system simply by adjusting the cycle time to the actual influent conditions at the expense of the time set aside for the idle phase. The idle phase thus serves to buffer peak influent loads. The consequence is that the react phase is kept in the range required to maintain control over the biological system (controlled unsteady state) (Wilderer et al., 2001).

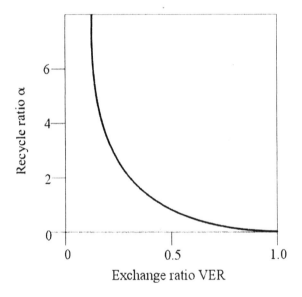

Figure 2. Comparison of volumetric exchange ratio in the activated sludge SBR with the recycle ratio in a continuous flow system.

In Figure 3, (after Wilderer et al., 2001) SBRs and continuous flow systems designed for denitrification are compared. In Figure (3b) a continuous flow denitrification system with one anoxic and one aerobic reactor are shown, which is translated into the SBR cycle shown in Figure (3a).

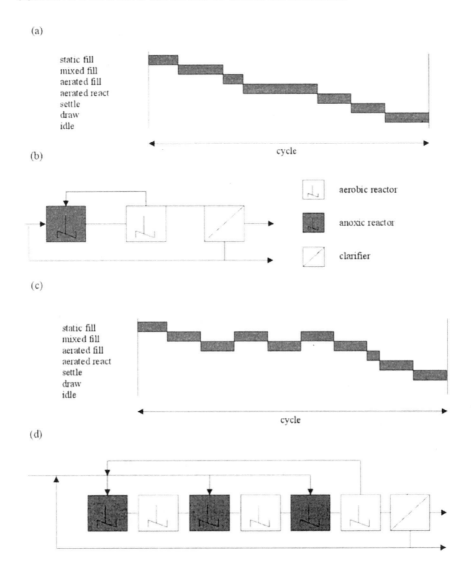

Figure 3. Comparison of SBR's and continuous flow systems designed for denitrification.

As can be seen from Figure 3b and 3d continuous flow systems bypass wastewater and recycle sludge to develop flexibility. Recycling is directed mostly to the first reactor. Bypass flow typically goes to anoxic or anaerobic tanks to supply electron donors for the removal of phosphorus and/or for denitrification. The equivalent action in an SBR is the application of aeration and mixing during react (except after static fill).

Pre-denitrification can be achieved in SBRs and continuous flow systems. The wastewater is introduced with only one change in mixing and aeration during fill (3a) or at one point in the continuous flow system (3b). The wastewater is introduced with mixing and aeration altering three times during fill and react (3c) or at three points of the continuous flow system (3d).

4.   LOWER OXYGEN DEMAND

The claim that SBR's have lower oxygen utilization has been repeated so many times but substantiation is rare. Sometimes credits is taken for the oxygen regained from denitrification. These same claims were made for the earlier extended aeration plants. However, oxygen consumption was still higher than that of the low sludge retention time (SRT) plants. High rates of simultaneous nitrification and danitrification were also observed. Nichols and Osborne (1975), (after Barnard, 2001) reported 85% nitrogen removal was observed in the extended aeration plants when about one third of the aerators were switched off. Substantial modelling which led to the development of the IAWQ model shows that configuration of the activated sludge plant will use the same amount of oxygen at the same SRT for carbonaceous removal when based on COD (Barnard, 2001). Since there is a wide range in the COD/BOD$_5$ ratios, depending on the hydrolization of the carbonaceous compounds in the wastewater, the oxygen uptake rate may vary from one plant to the next. Barnard, (2001) observed ranges from 1.5 to 2.5 (60% of differences). These differences are then attributed to a particular process.

5.   SLUDGE BULKING

One of the repeated claims of superiority of the SBR process is the production of non-bulking sludge. According to Silverstain (after Danesh et al., 1997) high fluctuation of supplied load do not affect the process course. No mixing during fill phase can save energy and limits growth of filamentous bacteria. Shaker et al., 1994 noticed considerable improvement of sedimentation property during anoxic filling compare to fill with aeration. Bulking in nutrient removal plants result mostly form the growth of Microthrix Parvicella, which relates to a low dissolved oxygen environment. It would appear that most of the full-scale SBR plants exhibit a high degree of bulking which can be explained by the need to operate at low DO values in order to effect simultaneous nitrification and denitrification (Bernard, 2001).

6.   PERFORMANCE IN TERMS OF NITROGEN REMOVAL

The main problem associated with nitrogen removal stems from the diurnal nature of the discharge. Total nitrogen load reaching the plant may peak at between 1.8 and 2.8 times the average load. Since the nitrifying organisms cannot store nitrogen and can treat only that which is immediately available, a high diurnal variation usually results in a peak of ammonia in the effluent, the severity of which depends on the peak to average ratio. For this reason a safety factor was normally assumed which was of the same order as the peak to average ratio (Banard, 2001). According to Jetech, 1992 (after Danesh et al., 1997) due to the large sludge mass in the aeration tanks, a safety factor for SBR technology was not necessary. Bernard and Sears, 1996 showed that contrary is true and that safety factors, or allowance for peaks of ammonia may be even more applicable to SBR's. Due to the short cycles used especially during wet cold weather, a proportionally higher load is introduced over a short time into one of the two or three units, resulting in unconverted ammonia in the influent.

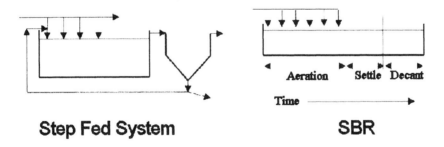

Step Fed System   **SBR**

Figure 4. Step Fed System and SBR used for nitrification and denitrfication

In the step-feed system used for nitrification and denitrification the feed is introduced along the length of the basins into anoxic ones formed at these entry points (Figure 4). This system has the advantage that a large portion of the mixed liquor is retained in the first number of sections, which allows the breakdown of adsorbed carbonaceous matter and ensures that the nitrifiers are not washed out of the system. However, since a portion of the influent organic carbon and ammonia enters near the end of the tank, some will wash through to the effluent without endangering the system. This will then produce a peak of ammonia in the effluent. The feed into an SBR is similar to the pattern in a step- feed plant and the results are similar. The only difference is that what happens in space in one instance happens in time in the other. It is therefore difficult to get consistently low effluent ammonia from a SBR unless the SRT is increased substantially (Barnard, 2001).

7.   CONCLUSIONS

It is widely acknowledged that SBR process is a feasible alternative to continuous-flow systems in carbon and nutrient removal. SBR systems offers a great deal of operation flexibility. It allows for easy adjustment of cycle time and HRT to variable inflow conditions. SBR system can easy operate ensuring low, constant concentration of substrate in reactor or alternate high/low concentration of substrate during fill phase. It gives a possibility keep to a minimum growth of filamentous bacteria and sludge bulking. The main stream of wastewater passes through the continuous flow plant by gravity, with bulk of the pumping being required for the recycling sludge. SBR systems do not require recycle because sedimentation takes place under quiescent conditions within the biological reactor itself.

Because of nutrient biological removal is controlled by sludge retention time and availability of volatile fatty acids (VFA) being organic carbon source, integrated removal of N and P in SBR systems is difficult. Phosphorus removal during anaerobic phase is inhibited by high nitrate concentration in the system and shortage of VFA. Proper adjustment of aerobic, anoxic and anaerobic period during the cycle and using of external VFA source are necessary to gain high efficiency for SBR systems.

Full-scale experience shows that SBR is not the most ideal process for biological removal except when favourable conditions exist.

ACKNOWLEDGEMENTS

This study was supported in part by the Kościuszko Foundation, An American Center for Polish Culture, Promoting Educational and Cultural Exchanges and Relations between the United States and Poland since 1925.

REFERENCES

1. Artan, N., Wilderer, P., Orhon, D., Morgenrotr, E., Özgür, N. The mechanism and design of sequencing batch reactor systems for nutrient removal – the state of the art. Wat. Sci. Tech. 43 (3), 2001, 77-84.
2. Barnard, J.L. Sequencing batch and flow through activated sludge systems – a fair comparison. 2001. Unpublished.
3. Danesh, S. Sears, J. Barnard, J.L., Oleszkiewicz, J.A. Biologiczne usuwanie związków biogennych w cyklicznych systemach SBR. International Conference „Usuwanie związków biogennych ze ścieków". Krakow, 1997, June 16-18, 1997.
4. Goronszy, M.C., Rigel, D. Simplified biological phosphorus removal in a fed-batch reactor without anoxic mixing sequences. J. W P C F 63 (3), 1991, 248-258.
5. Irvine, R.L., Ketchum, L.H. Jr . Sequencing batch reactor for biological treatment. CRC Crit. Rev. Envir. Control. 18 (4), 1989 ,255-294.
6. Sheker, R.E, Aris, R.M., Shieh, W.K. The effect of fill strategies on SBR performance under nitrogen deficiency and rich conditions. Wat. Sci. Technol., 28 (10), 1994, 259-266.
7. Teichgräber, B. Sekwencyjne reaktory porcjowe. Projektowanie i zastosowanie. Gaz Woda i Technika Sanitarna. 72 (12), 1998, 522-531.
8. Teichgräber B., Schreff D. Ekkerlein C., Wilderer P.A. SBR in Germany – an overview. Wat. Sci. Tech. 43 (3), 2001, 323-330.
9. Weber, W.G. Jr and DiGiano, F.A. Process dynamics and environmental systems. New York: John Willey and Sons, 1995.
10. Wilderer, P.A., Irvine, R.L., Goronszy, M.C. Sequencing batch reactor. IWA Publishing, 2001.

Index